Feeding Imaginations

Worship That Engages

Michael G. Bausch

Material in Chapter 1 appeared in the May 2008 issue of *Liturgy: Journal of the Liturgical Conference*.

Portions of the material in Chapters 2 and 7 appeared in my 1997 doctor of ministry thesis, "Using Video Resources in the Worship Setting," and in a series of articles in 1999 in *Church Worship: Resources for Innovative Worship*.

Material in Chapter 4 appeared in 2009 in *Liturgy: Journal of the Liturgical Conference*, and some data in this chapter's version have been updated to reflect research available in 2015.

Material in Chapter 5 appeared in 2008 in *The Clergy Journal*.

Material in Chapter 8 appeared in 2007 in *The Clergy Journal*.

Biblical references are taken from the *New Revised Standard Version Bible*, copyright 1989, by the Division of Christian Education of the National Council of Churches of Christ in the United States of America.

About the cover: The cover image is of the Ralph Ley Chapel at Pilgrim Center, Green Lake, Wisconsin. The three windows point to the Holy Trinity, with the cross in the central window recalling the story of the crucifixion of Jesus of Nazareth, and the two side windows representing the story of two others crucified on either side of him. Note a labyrinth in the floor, a central communion table with a smaller cross, a simple lectern/pulpit on the left side, and a small pillar of stones to the right. The sanctuary is built on the site of an old Native American summer encampment and honors the sacredness of the place, while looking out over the oak trees and the lake behind it.

ISBN 978-0-9864407-0-0

Table of Contents

Dedication

I continue to be grateful to those churches and institutions that nurtured, taught me, and enlisted my skills and gifts along the way:

Bethany Presbyterian Church, Milwaukee, Wisconsin
First Congregational Church, UCC, Watertown, Wisconsin
Carroll College, Waukesha, Wisconsin
Evangelical and Reformed UCC, Waukesha, Wisconsin
North Shore Presbyterian Church, Shorewood, Wisconsin
The Wisconsin Conference of the United Church of Christ, Madison, Wisconsin
United Church Camps, Inc. Green Lake and Moon Beach, Wisconsin
Pacific School of Religion, Berkeley, California
Arlington Community Church, Kensington, California
Montclair Presbyterian Church, Montclair, California
The Northern California Council of Churches, San Francisco
Wholly Mackeral Productions, Berkeley, California
The California-Nevada Conference of the United Methodist Church, San Francisco, California
The Northern California Conference of the United Church of Christ, San Francisco, California
The Interfaith Communications Commission, San Francisco, California
First Congregational United Church of Christ, Redwood City, California
United Church of Christ (Congregational), Williams Bay, Wisconsin
The Pilgrim Press, New York, New York
Union-Congregational Church UCC, Waupun, Wisconsin
The Center For Arts, Religion, and Education, Berkeley, California
United Theological Seminary of the Twin Cities, Minnesota
The Alban Institute, Bethesda, Maryland
The University of Dubuque Theological Seminary, Dubuque, Iowa
Summit Congregational Church UCC, Dubuque, Iowa

The Iowa Conference of the United Church of Christ, Des
    Moines, Iowa
Philosophy Department, University of Wisconsin-Platteville,
    Wisconsin
St. John's United Church of Christ, Hartford, Wisconsin
St. John's United Church of Christ, Slinger, Wisconsin

I add my thanks to Fred Noer of Image Source
(www.frednoer.com) for his editorial advice, the work he's done in
the editing and layout of this book, and for our many conversations
about photography and writing.

Thank you to my family: my grandparents Jacob and Florence
Bausch and Hazel and Armand Weiss, who laid a Protestant and
Catholic church foundation for me; my parents George and Marilyn
Bausch, who helped raise me in the church and who encouraged my
experiences and development; my wife Cathy for getting me to
world-class art museums and teaching me how to see and understand
art; to our daughters Anica and Brianna and their broadening my
musical and cinematic horizons; my son-in-law Aaron for his guiding
me to some really big fish; and to our grandchildren Livia Mae and
Wade Michael, who eagerly welcome and share in the art and music
projects we enjoy together!

Michael Bausch
January 2015

Preface

The reason for this little book is simple: I wanted to make available my previously published articles on the use of media in worship along with new material on how to preach the worship architecture of a holy place. The articles, updated as needed and set into the chapters you find here, and the new chapters, primarily focusing on how to preach the architecture, furnishings, and fabrics of worship sanctuaries, have grown out of significant experiences I shared with others who were similarly dedicated to creating worship experiences that truly made differences in people's lives.

The ideas and experiences detailed here have grown out of the way I and others worked together to find fresh new ways to help worshippers reflect on the meaning of their own lives in relation to the world around them as they respond to God's calling to be a blessing to "all the families of the earth" (Genesis 12: 1-3).

My previous book, *Silver Screen, Sacred Story: Using Multimedia in Worship* (The Alban Institute, 2002), describes the development of a weekly media-intensive worship service that went on for some years in a small Wisconsin community and the teams of people involved in envisioning the service, developing it, and presenting the experience.

After writing that book and then going on the road to share what we learned with churches, seminaries, clergy, and laypeople, I was asked by several publications to continue to reflect on what I was learning and to present that to their readers. Some of these chapters include these articles in their original form or updated to reflect current circumstances. These chapters share what I think are findings of great significance for those interested in taking their own use of media in worship to levels beyond simply putting hymn lyrics, prayer texts, or scriptural passages on a screen.

As I said in *Silver Screen, Sacred Story*, we live in a visual culture saturated with the language and screen devices of new media. I continue to see the benefits to adding a big screen to a worship space where the screen can capture and hold attention in fresh new ways and can stimulate spiritual imaginations among members of our congregations.

What is in short supply in culture, however, is conscious reflection on the content and meaning of all that we watch. This is something the church can model and practice with its use of film, photography, art, and literature to deliver biblical, theological, and ecclesial content and meaning.

In the congregations I have served since leaving the church in which I first did my visual experiments described in *Silver Screen, Sacred Story*, I introduced visuals without the long process I described in that book. In one congregation I served (2007-13) there was a large white wall to the left of the pulpit, perfect for projecting imagery. In my first few months there I put my personal projector (bought in 2006 for $500) on a media cart on wheels and connected the projector to my laptop, and with a remote mouse I could advance slides while delivering my illustrated sermon. I enlisted a few teenagers to develop visuals for announcements, prayers, illustrating hymns, and promoting youth projects for projection on the wall before and during the worship service.

This older congregation accepted the use of visuals immediately and offered lots of helpful feedback. One of the ninety-year-olds suggested we keep the image a few inches above the door that was below our projection area to make the image visible throughout the sanctuary. Another person asked me not to use the red dot laser pointer when pointing out an aspect of a picture during a sermon. She noticed I was moving the pointer quickly, and the dot was irritating her eyes, making her afraid of getting a migraine! I learned from this that some people with epilepsy find this uncomfortable as well. I stopped using the laser pointer. The ease with which we could get visuals on the wall next to the pulpit, to the congregation's right from the preacher (we read from left to right, from the speaker to the imagery, and back again) made using visuals a welcome and frequent mainstay in our worship services.

A very effective use of imagery served the historical nature of this particular church. We were able to photograph and document the installation of the old church bell on a new tower situated outside the sanctuary and then show the pictures over the months leading to the 120th anniversary celebration of the founding of the church. Photographs of the anniversary worship and dinner were shown before worship started in the following weeks to remind people of the wonderful celebration.

The first time I used visuals in worship at two churches I currently serve as interim minister (2013-present) was without a projector and screen: I simply studied the stained glass windows in the sanctuary and developed a couple of sermons about the windows, their stories, and their colors. This short sermon series not only captured everyone's attention, but also modeled a way of looking at religious art and architecture and drawing from it scriptural and theological lessons.

Shortly after that introduction to seeing and talking about art as a part of worship, I enlisted the confirmation class to illustrate the words to the song "I Am the Light of the World" by Jim Strathdee. In our class sessions the students loved singing the song, so I asked them to start drawing illustrations for each phrase in each verse. We photographed those pictures with a smartphone and then uploaded them into a Google Docs presentation, which we edited together later in the week via Gmail.

On the Sunday we presented this song (at both churches), I put my portable projector on the communion table at the steps of the chancel and put a large Bible in front of the projector to block its view from the congregation. I attached a laptop computer to the projector, with a cord long enough so one of the sixth grade boys could advance the slides from the front pew as we sang "I Am The Light of the World" accompanied by the youth who provided the imagery and me on guitar. For a screen we used an old portable tripod screen on loan from a member's workplace. We positioned the screen in the center of the chancel and removed the screen once the song and illustrated sermon were finished so as not to block the altar and cross.

Following the song I preached a sermon on light using a number of pictures. At one church I noticed a longtime member in his 70s do a double take when he saw pictures on the screen: His eyes were "hungry," and he was instantly attracted to the large visual in front of the sanctuary. It's a great feeling for me to see the eager attention people offer to what we show on a screen in a worship setting that hasn't used screens much – if at all!

After this worship service the chair of the Pastoral Search Committee thanked me for this introduction to the use of visuals in preaching and worship because, he said, his committee had noticed applicants for the pastoral position had mentioned their skills and interests in using visuals in worship.

This book, then, grows out of my experiences with people in congregations like these who yearn for spiritual growth and who continue to welcome the ways worship can feed the imaginations of people while also inviting them to share their inspired imaginations with others.

I continue to share the optimism of Old Testament professor Walter Brueggemann who believes "the purpose of preaching and worship is transformation." He defines this transformation as "the slow, steady process of inviting each other into a counterstory about God, world, neighbor, and self." This is accomplished through liturgy and proclamation that recognize "people in fact change by the offer of new models, images, and pictures of how the pieces of life fit together – models, images and pictures that characteristically have the particularity of narrative to carry them." [1]

The narrative of worship is the story of faith as found in scripture and interpreted by tradition and the gathered community today. Each time we gather to worship, we are asked to experience the narrative once again, by means of everything worship leaders use to communicate the story that day. By feeding imaginations and engaging hearts and minds, be it through story, song, calling attention to the art and architecture of the very room in which people gather, or by inviting in the world with a computer, projector, and screen, we continue to "tell the old, old story."

[1] See Walter Brueggemann, *Texts Under Negotiation* (Minneapolis: Fortress Press, 1993), pp. 24-25.

Imagination is more important than knowledge. For knowledge is limited to all we now know and understand, while imagination embraces the entire world, and all there ever will be to know and understand.

– Albert Einstein

Introduction

How does a church design worship that reaches the iPod, cellphone, digital camera, video clip, screens-everywhere generation of teenagers and young adults? How are visuals incorporated into worship, and what kind of music will best relate? How does a "traditional" church grow to accept and support a new worship experience designed for youth, young adults, and tuned-in baby boomers? What technologies are required, and what resistances are to be expected? How does learning-theory research impact what technologies are used? What theological basis is there for a movement away from print-oriented worship to electronic-oriented worship?

These are some of the questions many church leaders, lay and clergy, and more and more congregations are beginning to address as they look to provide meaningful worship experiences for a generation of people who are both electronically oriented and spiritually hungry.

Almost twenty years ago (how time flies!), church growth consultant Bill Easum wrote, "Nothing is undergoing as much change in the life of a thriving church as is worship. This change, much more than just radical, is revolutionary. The forces that have driven worship for five hundred years – the printed page, a 16th-century appreciation of music, and a culture that embraced Christianity – are being replaced by MTV, the Internet, multimedia, and a post-secular culture. The day of the hymnal, overhead projectors, organs, and processionals is drawing to a close. The day of visual, virtual, multi-user, interactive worship is upon us." [1]

> " . . . may your art help to affirm the true beauty which, as a glimmer of the Spirit of God, will transfigure matter, opening the human soul to the sense of the eternal."
>
> John Paul II

More and more churches are using video in worship to project hymn lyrics, to visually illustrate sermons and announcements, and to communicate more effectively with a video-clip generation. The articles contained in this resource are written to provide the practical information useful to laity, clergy, and churches interested in adding the visual arts to worship settings. The articles suggest how worship communities initiate, organize, develop, and critique a commitment to the communication of the word of God through the visual arts and other media. This book also discusses positive and negative aspects that surround our use of technology and builds an imaginative theological justification for using screens – and all that they may display – in worship.

While many of us think worship screens are a "thing" of recent years, what most do not know is that churches have been using screens and projectors in worship since the early 1900s. Recent research has uncovered how Congregational ministers as early as 1910 were using film during worship services to "drive home the message" and to put the Gospel into "pictorial form." At the turn of the 20th century, Thomas Edison wrote for *The Congregationalist* and an audience of laity and clergy how their churches could begin to use motion pictures for "instruction and moral advance."

Since the 1930s, Roman Catholic popes and bishops have advocated developing an awareness of cinema, its power, and effects and discovering how the medium could be used to fulfill the mission of the church. In 1995, John Paul II wrote, "Among the media of social communication, the cinema is by now a universal and esteemed medium from which messages are often sent which are capable of influencing and conditioning the choices of the public, and especially young people, in a form of communication that is based not so much on words as on concrete events, expressed in images which impact greatly on the viewers and on their subconscious. The cinema, since it was invented, while sometimes giving rise to criticism and disapproval on the part of the church on account of some aspects of its extensive output, has also often dealt with themes of great meaning and value from an ethical and spiritual point of view." [2]

Embracing – and resisting – the electronic media arts, technologies, and messages within the context of Christian mission, education, and worship is as old as electricity – and as new as the latest digital technological innovations!

Churches that have used screens in worship, and that are investigating ways to incorporate screens even more, understand the interplay of these important factors:

• The church's attentiveness to its mission of evangelism and effective teaching, illustrated by churches using video film clips and other projected media as resources for worship, and assertions that "the fastest-growing congregations in the nation, whether conservative or liberal, use visuals in their liturgies." (The late Doug Adams, professor at Pacific School of Religion, Berkeley, California)

• The church responding both to research pointing to high learning retention when hearing is combined with seeing (some studies indicate retention nearly triples when listeners see as well as hear) and to advertising efforts by large technology companies promoting their multimedia products as effective teaching tools.

• Church leaders interested in incorporating technological advances into worship and learning settings. For example, new church building designs include architectural configurations that allow for large projection screens and the use of multimedia/video technologies.

• The widespread cultural acceptance of seeing a variety of video screens in varied settings. Most people are comfortable using visual technologies in home, work, and school and are increasingly reliant upon them for learning and entertainment purposes.

What follows in this book is material that has grown out of years of experience with film, music, photography and art; of years of intensive group process, experimentation, and creative liturgical and preaching practice within the "laboratory" of several worshipping congregations; of several years of writing articles and a book about these experiences; and of years of teaching workshops, seminars, online courses, and summer sessions with clergy and laity dedicated to worship innovation.

This book, then, is intended to be a resource for those:

• wanting to learn more about the practical uses of screens, visual arts, popular culture, projectors, and computers in worship.

• willing to think through the positive and negative implications of the use of such resources.

• committed to the ongoing creative enterprise of stimulating and engaging the imaginations of those who gather in worship communities.

Part I

Engaging

Epworth United Methodist Church, Berkeley, California

These first chapters of the book share what I've learned about the power of visuals in worship and how these stimulate imaginations and open doorways to individual and congregational hearts.

Chapter 1

Celebrating the Possibilities:
Media Art in Preaching and Worship

The perspective I bring to this conversation is as a parish pastor with a long experience of using visuals and visual technologies in worship and preaching. During the 1960s and 1970s many of us tried using 35mm slides, record albums, and videotapes to offer what is now called "liturgical media art" in worship and education. Somewhere in the middle of the 1990s technologies changed, and it became even easier to bring pictures, music, and video into sanctuaries. During this time I developed, along with teams of laity, a weekly media-intensive worship experience in a mid-sized Midwestern church.

During those years I also began to offer workshops on the use of media in worship to denominational gatherings, taught summer courses at a number of mainline seminaries, helped develop a doctor of ministry program in digital media and congregational revitalization, and wrote a number of articles and a book about screens, media art, and worship. All of this is to say that while I have long experience using media in preaching and worship, I have combined this with research in the field to better guide clergy and laity also interested in developing or improving liturgical media art in their worship and preaching settings.

It is not my purpose to define for you worship, preaching, and liturgy. I trust you have your own understanding of their origin, purpose, function, and presentation as they happen each week within the life of your own worshipping congregation. Generally speaking, many of us would agree what happens in worship is people gather to:

Come into God's presence, Engage the Word of God, Respond to the experience of that Word, and Be sent back out into the world.

Much of how this is accomplished is in telling the stories of faith. This is done through storytelling in many dimensions: the announcements of the congregation, where people are told what has happened and what is to happen in the life of the gathered church; the singing of hymns, songs, and anthems in which stories, scriptures, and theological lessons are presented; the liturgies of prayer in which life stories, plots, tensions, and resolutions are named; the reading of scriptures through which biblical stories are remembered; and the preaching of sermons filled with illustrations, anecdotes, and stories organized to draw meaning from the scriptures.

Most of what I have described is familiar to worship leaders and follows the structure of how this storytelling occurred in biblical times. The song of Miriam at the sea in Exodus 15:20-21 is a celebration and a testimony to the story of the deliverance of the Hebrews from the Egyptian oppression. Joshua called for 12 stones to be carried across the Jordan River and placed on the promised land (Joshua 4:4-7) so children will see them and ask, "What do these stones mean?" in order that the story can be told. Jesus told parables, which then raised questions in his listeners' minds and stimulated further discussion for the sake of a life lesson.

Songs, monuments, and parables all are means to tell the stories of faith. There is yet another way to tell stories, and it was used by the early churches when they painted simple pictures and symbols in the catacombs and by later congregations as they added mosaics, stained-glass windows, painted canvases, and fabric art to show the stories and their symbols in the worship sanctuaries. These media, put together into a harmonious whole (which is the art and science of rhetoric), helped tell the stories of faith effectively while inviting a worship congregation to remember these stories, apply the lessons learned to the congregants' own lives, and experience the catharsis of being grasped and pulled in by the story.

Today's presentation technologies make it easy to continue this long tradition of storytelling in worship while increasing the options for doing so. As I have learned while developing weekly media-intensive worship services for eight years, this can bring changes to a worship community. What primarily started as a means of telling the stories of faith more effectively to a visual (and audiovisual) generation in worship by picture and parable from popular media, came to bring many unanticipated results.

Increases in Worship Attendance

As with all increases in worship attendance, the increase was due to a number of factors. While the main factor for this was adding an additional worship service to the Sunday morning schedule, making the service visually centered and culturally relevant created an interest among those searching for something new and different. People started coming with the expectation that something new, different, and engaging would be presented as part of the worship service. More young adults and families with children started attending worship more regularly. They represented a generation accustomed to screens, relevant music, and film clips, but for whom the application of these materials to biblical story and theological theme became a draw. More men began to attend worship, partly as a result of the simplification of the liturgy by reducing the number of words that were read, increasing the number of pictures that were seen, and by reducing the number of hymns that were sung. These increases also were a result of members inviting friends and family into the worship service. The media service created a new sense of evangelism. People who felt the multimedia worship experience met their needs through its attention-getting excitement and cultural relevancy became very comfortable inviting friends and acquaintance to this unique experience.

Increases in Mission Giving

We came to see how effective visuals were for helping people better understand the mission causes the church was promoting through special offerings. Pictures of these appeals – and in some cases brief video clips – helped draw people into the story of the project and give to support these very specific needs. Thousands of dollars were raised for victims of Hurricane Katrina because we were able to show pictures of local people on site in Louisiana purchasing and delivering diapers, towels, and blankets from the initial donations provided by our church. More gifts followed these pictorial reports. These e-mailed pictures helped dramatize the urgency of the situation as well as show the specific ways our congregation's donations were being used. In another instance, hundreds of dollars were quickly donated to help a local family with expenses associated with the premature birth of twins and their hospitalization in a nearby city. We were able to show pictures of the newborns on a weekly basis and report their progress as well as highlight the amounts being donated by church members toward various medical needs. Having the projector, screen, and computer in place in the sanctuary made it easy to "think visually" and show the various causes we were supporting, thus making it easy to raise much-needed funds.

Increased Engagement With the Sermon

Another response to the use of visuals in worship was the increased discussion and interaction with material presented during "visual sermons." With the addition of illustrations to a message, including symbols, diagrams, film clips, and photography, people could enter a message at different points stimulated by the visuals, reflect upon what was seen and heard through the persons' own experiences, and gain a memorable experience to talk about afterward with others. Pictures made the preaching themes more memorable by adding an additional sense of the visual to the auditory. Besides the stimulation of dialogue about sermon presentations, there seemed to be more catharsis resulting from some of the powerful visual illustrations. We noticed there were more tears in worship as a result of music combined with rich imagery and linked with scriptural text and theological theme. Dramatic and emotionally gripping film sequences added to the catharsis in unexpected ways. People were touched and drawn more deeply into the sermon message through the use of visuals and audio in the sermon.

An Expanded Opportunity for Preaching Contemporary Issues

Those clergy wishing to preach contemporary social issues can find an easy alliance with visual media. All of today's social issues are communicated through the popular media of film, television programs, books, newspaper and magazine articles and photographs, and in many songs. Preachers can select from a wide variety of visual resources to illustrate a topic and present an outline for understanding the issues involved as well as to demonstrate a theological viewpoint. In their preaching and during worship, church leaders can use the screen to more effectively communicate and promote theological, political, economic, and social issues. Poverty, war, racism, gender inequalities, consumerism, and global warming are easily illustrated with photograph, art, film clip, and animated graphics as part of a sermon illustration, prayer litany, or invitation to the offertory. Preachers and liturgists can visually show local, regional, national, and global church efforts to reduce the effects of poverty, to protest war, to dismantle racism, to foster cross-cultural communication and competencies, to educate about HIV/AIDS, to eliminate the death penalty, to mobilize efforts to face up to global warming and encourage necessary lifestyle changes, and to affirm a diversity of loving relationships and family systems.

As the Bush Administration developed its case to invade Iraq, we offered sermons about nonviolence as a Christian approach to international conflict. We found pictures to accompany the U-2 song "Peace on Earth" and showed Iraqi families in their homes and communities. Through this multimedia approach we hoped to put a human face on the Iraqi people. As part of the song we posted on the screen quotations from Gandhi, Martin Luther King Jr., Nietzsche, and Jesus about nonviolence. We found old newspaper articles from the local community during the Vietnam War where people questioned the role of God in times of warfare.

> "Art has a unique capacity to take one or other facet of the message and translate it into colours, shapes, and sounds which nourish the intuition of those who look or listen. It does so without emptying the message itself of its transcendent value and its aura of mystery."
>
> John Paul II

As our state was considering a constitutional amendment to define marriage as solely between one man and one woman, we provided a Valentine's Day sermon that featured illustrations from traditions surrounding the original St. Valentine, who was imprisoned by the Roman government for illegally performing marriages of soldiers who were not to be married while they served Rome. His was an example of affirming loving relationships even when the government had passed laws defining who could marry and when.

On a Memorial Day weekend we juxtaposed patriotic songs about America with film clips from the film, "The Fog of War" (2003), featuring video and audiotapes of Lyndon Johnson and Robert McNamara willfully deceiving the American public about sending more troops to Vietnam. These media vividly illustrated the vigilance people must maintain when governments wage war without much public accountability.

After a local clergy association expelled Muslim and Wicca members of the group (these clergy were chaplains in a state correctional facility in the community), we outlined the necessity of interfaith dialogue through on-screen images of newspaper headlines and articles about the local controversy. Scenes were shown from a PBS documentary about a similar controversy that was generated when a Lutheran pastor prayed with people of other faiths at a September 11 memorial service at Yankee Stadium.

On Mother's Day we showed photographs of a local dog who had adopted orphaned kittens and then used a song about a nontraditional family composed of two lesbians raising a child together, showing pictures of a diversity of family configurations, including multiracial and same-gender families, and more pictures of that dog mothering kittens! This was to illustrate a message about the wide embrace and compassion we experience from many mothers and how this is a call to all of us that comes from Christian faith.

In all of these cases the screen was effectively used to show the issues being discussed through the use of photographs, artwork, quotations, and film clips. The screen became a new window in the sanctuary, a window where, with imagination, creativity, and passion, a media arts team can boost the preacher's persuasive appeals from the pulpit.

Widespread Involvement by Members of the Congregation in Planning and Presenting Worship

As much of our screened material was locally produced, this approach created the need for more people to become involved. As people caught the excitement of what happened by adding visuals to worship, some of the persons learned how to share their own artistic sensibilities by combining imagery with music or finding film clips of scenes relevant to theological discussion and preaching themes. While some people came forward through general invitations, others stepped forward on their own, volunteering to assist in small, manageable projects. Those who were already involved and who had shared certain experiences could help bring others in and train them.

Further, this called attention to the fact that many people in the congregation had skills that were underused by the church. By adding visual arts to worship we invited in the photographers, graphic artists, art educators, artists, videographers, and latent media artists to share their gifts to illustrate and present various preaching and theological messages. This in turn served to reactivate inactive members. We found that the typical overreliance upon musical arts in worship brings in musicians and singers of a congregation yet excludes many who are visually oriented. Some of them are nominally involved in church because the church hasn't met their needs. Such persons return when invited back in to share and experience worship with a visually rich environment. This participation came to affirm the continued use of visuals in worship. Such positive reactions grew a culture of expectation the church would continue to use the screen in direct relationship to preaching and worship.

As youth and adults became comfortable developing PowerPoint presentations with original photography and appropriate music, a kind of cottage industry developed that provided limited opportunities for part-time work. As more people caught on to the power of visual displays in special events such as weddings, funerals, and special worship occasions such as confirmations and celebrations of graduations, church members were asked who knew how to do this to provide people with media art for these family occasions. Families needing wedding and funeral presentations hired young people with their skills at camera work, video editing, and scanning and creating slide shows. This arrangement provided a source of income for teens and adults who were able to develop the media art that fit these occasions.

Most Important Change: The Blossoming of An Interpretive Community

Probably the most important unexpected change that took place as a result of the introduction of media arts in worship was the development of an interpretive community. As more and more people became engaged directly with creating "liturgical media art," they came to understand the importance of having theological conversations to link biblical text with cultural media productions such as film clips and pairing of images to music and lyrics. Rather than becoming recipients of someone else's messages about scripture during worship, the people became direct participants with a stake in a creative engagement and application of scripture to cultural media. People developed a fluency with a kind of hybrid worship language from the creative interchange between cultural media and theology.

The interpretive community became widespread. Large numbers of people became involved in many different aspects of developing, preparing, and presenting liturgical media arts or in becoming part of the milieu in which the media were nurtured and shared. Small groups met regularly to study lectionary texts, select themes, and brainstorm film clips and music that could be used to help illustrate those themes. The core team invited others in the church to find pictures, art, music, and lyrics that seemed to connect with various biblical themes and stories. Some watched films and found scenes with obvious connections to "churchy" topics such as prayer, forgiveness, sin, peace, love, and justice. Those who joined in these efforts developed the practice of drawing preaching themes out of cultural "texts" or finding "texts" to support preaching themes. Some people grew fluent in this process and suggested or developed entire worship services that focused on a theme of personal interest that could be easily illustrated by film and music. A group of nurses, for example, prepared a worship service around the relationship of health and faith.

Film groups, book groups, and bible study groups all fed into the experience of weekly worship. Photographers and painters saw the connection of their arts with worship. Trips to area art museums and international travel changed the members' views of art, and new suggestions arose for using visuals in worship. Some members enjoyed the solitary experiences of working with computers to artistically enhance and edit photographs and videos to provide original illustrative material.

This interpretive community extended to those whose involvement was simply a part of their attendance at weekly worship. Some worshippers reported finding themselves making connections with Sunday's message and popular music and media they encountered the following week. The media triggered thoughts of the message that had been illustrated by the media the previous Sunday.

As a result of these surprising and often unanticipated results of using liturgical media art in worship, I came to understand and develop what I found to be three principles of interpretive communities. These principles help answer the question why the blossoming of interpretive communities matters in our churches.

Interpretive Communities Center Themselves
Around the Scriptures

Interpretive communities grow in relation to their mutual study of scripture. They find ways to experience scriptural story and theological theme through commentaries, poems, novels, paintings, films, short stories, sculptures, fabric art, movement, and dance. The communities expose their faith to the arts in all of their forms, and the conversations together create meaning through theological reflection. These communities come to understand their life and faith as lived in the midst of a particular location and community.

Interpretive Communities Equip One Another to Grow
in Understanding and Participation

Interpretive communities equip one another to construct, present, and interpret the messages of Christian faith through their educational, social, worship, and mission practice. Communities go to art museums and talk together about what they see; they attend concerts and films to experience, engage, and discuss with hearts of faith; they form book groups to read anything that informs hearts and minds about life, love, theology, critical social issues, or recent biblical scholarship; they select social projects and join efforts to identify systemic injustice, oppose war, eradicate poverty, dismantle racism, and work for gender equity; and they develop, present, and evaluate worship experiences that build community, honor and worship God, and send people equipped with faith and purpose out into the world.

Interpretive Communities Grow Over Time

All who join with a worship community participate in that particular community's history, tradition, and practice. This is a living process that evolves over a period. People have come and gone in the life of that community, and the community is always changing through births, marriages, confirmations, and deaths as well as through participation of the community in its entire social, educational, mission, and worship experience. People grow comfortable in these communities as time is spent with them. These communities of interpretation actively engage, interpret, and construct the meanings a broken world desperately needs. As the communities join together to create and support artistic interpretations of biblical story and theology, the members grow in trusting one another, develop means for whole-brain learning through the use of multisensory media, and formulate ways to engage the world through faithful witness.

Today's digital media technologies of screens, projectors, computers, and music playback devices make it easy and affordable for church members to create, store, and present attention-getting audio and visual resources for worship and preaching. In the hands of interpretive communities of faith, these technologies can contribute to a vibrant theological conversation that has the potential to make a transformative difference in the lives of those who participate. As my own experience has shown, these new media offer churches limitless creative possibilities for telling the stories of faith in engaging and memorable ways.

Interpretive Communities Need A Ready Supply of Resources

Listed here are three Web sites that continue to prove to be useful for creative worship preparation in interpretive communities. One site develops lectionary-based resources; one engages film, music, and other products of popular culture from a spiritual perspective; and one provides a theoretical foundation for understanding how to use today's media to reach a digital media generation.

http://www.textweek.com

This site is well known to many clergy as a starting point for weekly sermon preparation. While organized by lectionary text, the site offers the capacity to search for specific scripture passages, making it helpful to those not particularly bound to using the lectionary. The site is full of resources, including sample sermons, illustrations, stories, primary sources, and bible translations.

Of particular interest to those using projection technologies are the art and movie concordances. Works of art are searchable by their connections to scripture passages and themes and are presented in chronological order with the oldest art listed at the top of the page and the newest works at the bottom. This information is helpful for those looking for either more classical or more contemporary art, including that of artists from around the world.

The movie concordance also lists films and scene descriptions by their connections to lectionary texts and themes. Unfortunately, with most descriptions there is little guidance for where to find the scene in the film. The reader is left with watching the film, finding the scene, and previewing the content for its suitability in one's worship setting. While this is not the best solution for those hoping for a reference to quickly found DVD chapters and minute marks, textweek.com still provides a stimulating starting point for finding film clips appropriate to lectionary scripture passages.

http://www.hollywoodjesus.com/

This site affirms the presence of the gospel in today's films. Using the slogan "Pop culture with a spiritual point of view," the site provides reviews of films currently showing at theaters as well as DVDs available for rental and purchase. Each review offers a synopsis of the film's story line and adds commentaries from Christian writers sharing their perspectives and uncovering scriptural references, theological themes, and "what to look for."

I find the site very useful when I have a hunch there might be sermon material in a film I am either looking to rent or going to the theater to see. For example, with the entire "media buzz" surrounding a film such as "The Da Vinci Code," HJ offered plenty of study resources and interpretive guidance. That said, the site is more than a film database. Dozens of reviewers discuss current TV shows, music, fiction and nonfiction books, and comic books, all with an eye toward their spiritual contributions. You can stay current by reading blogs, listening to podcasts, or getting RSS feeds sent to your e-mail address. HJ is a great starting point for church leaders who wish to understand and to reflect upon the many messages communicated through today's popular culture.

http://www.marcprensky.com/

This site is helpful to those wishing to understand the impact of digital technologies on children and youth today and how teachers (and preachers) need to find new ways to communicate with "digital natives" of today.

Marc Prensky is an educator who creates video game-based training tools designed to teach technically fluent children, youth, and young adults. In his groundbreaking article, "Digital Natives, Digital Immigrants," Prensky coins the term "digital natives" to refer to those who are natural users of computers, video games, and the Internet. Those people not born into this digital world he calls "digital immigrants." Although they are using much of today's technology, digital immigrants retain an "accent" because they were born before the advent of home computers, cell phones, and the Internet. Prensky maintains that today's educational challenge is for the digital immigrants teaching in classrooms (and preaching in our churches) to find ways to effectively reach out to the "natives" in our midst.

At his Web site Prensky offers downloadable versions of his many articles about today's youth, how they think differently, and how their brains are changed as a result of their use of digital technologies. When navigating his site, click on "Writings" to go to a number of his free articles, including those with practical suggestions for developing effective teaching and learning strategies.

Prensky has given us a way to understand the shift that has been taking place ever since the development of the radio, the camera, the moving picture, and television: Those who use these media are affected by them. Prensky discusses how the "digital" world of technology is changing human brains. If brains are being changed and if multiple generations of people are shaped by electronically delivered content, how does the church harness this force for the sake of the gospel?

In the next chapter, we'll consider the importance of nurturing congregations that expect something to happen in their worship experience and how calling attention to visual arts may well bring a revelation of the Holy.

Chapter 2

Worship: Expecting Something to Happen

Psalm 131 is a song people sang on their way to the temple in Jerusalem. As they climbed the hill, some struggling in the hot sun, they sang, and in the singing they anticipated the holy place and the holy encounter with God:

"O Lord, my heart is not lifted up,
my eyes are not raised too high; I do not occupy myself with
things
too great and too marvelous for me.
But I have calmed and quieted my soul,
like a weaned child with its mother;
my soul is like the weaned child that is with me.
O Israel, hope in the Lord
from this time on
and forevermore."

Seeing this scripture again, we see that worshipping takes effort. It's not a form of entertainment where we just sit back, and when we leave we say, "Oh, wasn't that music beautiful?" Or, "Oh, wasn't that sermon exquisite?"

When we come to church to worship, what happens there happens best when we come expecting something to happen. You have to bring your self in. You can't leave your self at home and come to worship in body only. We have to come – ready to give it our full attention, ready to participate fully as we can, ready to notice God.

Worship is not something that happens around us or on the outside of us . . . it is something that has to happen on the inside. We have to prepare, make an effort; we have to be expectant and observant.

People in ancient Greece traveled to a temple at Delphi to speak with the Oracle. The Oracle lived in the Temple and served as the mouthpiece of the great god Apollo. People went to Delphi to bring their questions about their families, questions about personal problems, and questions about national and international crises. Those who came were looking for guidance, not answers, because the Oracle never gave clear instructions. Those who came took a long trip to the temple nestled amid the beautiful Greek hills overlooking the Aegean Sea. On this journey, it is said the people coming with their questions would think about the god Apollo and all of his attributes – light, insight, radiance, healing, order, elegance, and beauty – and would begin to anticipate the encounter they were going to have.

Rollo May, in his description of this anticipation, wrote how the worshipper's " . . . conscious intentions and his deeper intentionality would be already committed to the event about to take place. For the one who participates in them, symbols and myths carry their own healing power." [1]

The anticipation of the people going to Delphi would help them develop a commitment to the event that was to take place. Anticipation grew into an expectation that something would happen. Those who sought guidance brought along their hopes – and their faith. When each person asked the question that plagued his or her heart, he or she received a cryptic message from the Oracle, something requiring the individual to do his or her own thinking and interpreting. The Oracle never gave a clear answer. Each person who came to the Oracle was expected to freely respond and interpret the message received. Each person had to put himself or herself wholeheartedly into the encounter. These were not passive viewers and listeners but active participants. The questioner heard the advice and then was called upon to think fresh about his or her situation, using every creative and imaginative bone in his or her body.

This process describes what happens in worship. We must come expectantly. We must come ready to be actively involved. We must come ready to uncover the treasures we seek, knowing that God expects us to put an effort into it. Worship can be an hour of special alertness, where we come ready to notice a new detail, something we haven't seen before, or by noticing a feeling in one's self or the stirring of a new idea that grows out of some word or phrase that is heard.

If we come expecting it to be the same old and familiar, we will be the same old, familiar self when we leave. Is that the fault of the church or the preacher? Some might say so, but worship requires the worshipper's active participation. Each worshipper is a weaned child of God (Ps. 131), and that means each is no longer spoon-fed but knows how to feed himself or herself. God expects maturity in faith and willingness to grow and change toward what St. Paul called "full manhood" or "full womanhood."

When we come to worship, everything is in place for an encounter with God:

- Sacred space, dedicated to worship
- Architecture and symbol
- A holy time, dedicated to ritual and tradition
- Music and singing
- Opportunities for heart and mind to be engaged with God
- A relationship with a community of faith

But none of them make a difference without the individual worshipper.

We have to encounter what is presented. We set aside the time. We set aside normal life – to wash and dress and put on special "Sunday clothes" that are set apart for the occasion, to sing, to pray, to read aloud, to listen actively, to pay attention.

We might take a cue from the ancient Greeks and come to the place of worship with a question. For example, we might bring a question about personal matters, family matters, community life, or national and international issues. Anything and everything that concern us are fair to lift up to God for guidance.

The guidance that is received will cause us to think. We have to bear responsibility. God doesn't hand over the answer in a vacuum, but comes only to those who are receptive, who are waiting, who are developing their spiritual awareness, their religious sensibility, and their conscience of faith.

Worship, then, becomes a dynamic interchange between the ones who bring the questions and the God to whom it is presented.

The Oracle at Delphi provided a kind of mediation between the worshipper and the god, and today's worship leader is in a similar situation. Unless a congregation is asked to come with specific questions, however, a worship leader makes assumptions about what these questions might be. That being the case, adding visuals to worship can serve a congregation's desire to worship and to experience a revelation of the Holy One.

God is at the center of our worship practice. In worship, we gather in awe of God's love, justice, and creative power, and we offer thanksgiving, praise, and devoted gifts to God. In our worship we name our relationships with God, with each other, and with ourselves. Paul Tillich wrote that "Religion opens up the depth of man's [sic] spiritual life which is usually covered by the dust of our daily life and the noise of our secular work. It gives us the experience of the Holy, of something which is untouchable, awe-inspiring, an ultimate meaning, the source of ultimate courage."

The imagery of our words and our visual images are in the service of this revealing of God's grace, God's love, and God's relationship to our lives. In 1 Corinthians 14:26, we find Paul's characterization of "orderly worship" to include "a revelation." The Greek *apokalupsis*, "to remove the covering veil," finds its way into the Latin re-velum, "to turn back the veil," and into our English "reveal."

Tillich's discussion of revelation begins his *Systematic Theology*: "Revelation is the manifestation of what concerns us ultimately." This revelation needs "the word as a medium of revelation." Tillich is careful to define this "word." It is not to be narrowly defined as spoken, written, or heard words but is to accommodate " . . . the religious symbolism . . . which uses seeing, feeling, and tasting as often as hearing in describing the experience of the divine presence . . . the divine 'Word' can be seen and tasted as well as heard."

This is important as we develop a case for the use of visuals, including electronically mediated visuals, in the service of God's revelation in worship. Revelation is known through all of the human senses.

Theology disclosed and revealed through the senses is prominent in biblical narrative. The ancient *Shema* has served as a central affirmation of Israel's faith: "Hear O Israel: The Lord is our God, the Lord alone. You shall love the Lord your God with all your heart, and with all your soul, and with all your might." (Deut. 6:4-5) Israel was not only to "hear," but it was to write those words on the doorposts of their homes and recite and talk about the words with its children. The *Shema* served as an ancient "audiovisual" where hearing and seeing were combined with disciplined reflection and discussion.

In Exodus 3, we find Moses leading his father-in-law's flock to pasture. Suddenly a bush bursts into flame, and he sees the bush is not consumed. His full attention is captured. He says, "I must turn aside and look at this great sight . . . ." (Ex. 3:3) Then the voice of God calls to Moses from out of the bush, and he not only is engaged in seeing and hearing but responds verbally, takes off his shoes, hides his face, and feels fear. God gains the full attention of Moses in that moment of revelation through his visual, aural, tactile, and emotional senses.

Our word "attention" comes from the Latin *tendere* meaning "to stretch." To give attention is to "stretch toward." To give our attention is a physical stretching process of engaging all of our senses. Like Moses at the bush, we need our senses engaged in order to turn aside, stretching to become closer to God's revelation in a historical moment. God's revelation to Moses in Exodus 3 occurs through the light of a burning bush and the sound of a voice. All of Moses' senses are engaged in the encounter.

The prophet Isaiah provides another excellent example of theology disclosed and revealed through the senses. As an oral teacher, Isaiah uses sensory language and aphorisms because they maintain the listener's interest. Isaiah uses language that engages the senses and refers to that which is therefore familiar to the listeners.

Through 66 chapters, Isaiah employs references to the senses and the human sensory equipment:

    – references to hearing and the ear are made 87 times

    – speaking and saying, 245 times

    – the mouth, 27 times

    – seeing and the eye, 95 times

    – hands, 76 times

    – the feet, 19 times

    – the tongue, 2 times (it's mentioned only nine times in the entire Bible) and the tongue licking, 1 time

    – the experience of travail or labor at childbirth, 7 times

    – the nose, 4 times

Isaiah uses sensory language to capture the fullest attention and response of his audience.

Similarly, we find sensory communication in specific stories from Jesus' life. For example, when Jesus gathered with the disciples in the upper room, they were eating together. Taste and smell were added to the visuals of the meal, the table, the gathered friends, the sounds of talking and eating, and the feelings of the moment. Paul gives us the first account of that night, how Jesus " . . . took a loaf of bread, and when he had given thanks, he broke it and said, 'This is my body that is for you. Do this in remembrance of me.' In the same way he took the cup also, after supper, saying, 'This cup is the new covenant in my blood. Do this, as often as you drink it, in remembrance of me.'" (1 Cor. 11:23-25)

In speaking over the breaking of bread and the sharing of wine, Jesus engages the full sensory attention of his disciples and offers the truth of the Psalmist, "O taste and see that the Lord is good." (Psalms 34:8)

The disciples saw the bread and the cup and heard the words of Jesus. The disciples' feelings and thoughts were engaged by his telling the men to rethink the meaning of the bread and cup and to remember him from that moment on. The Sacrament of Communion was instituted using common, everyday material for divine purpose.

These examples illustrate the importance of engaging more senses in our communication. Biblical narratives demonstrate how hearer and reader are invited into a new reality through language and symbol.

> We see today a "shift away from a basic orality in theology . . . to a multimedia theology in which the almost total communication ambitioned in electronic technologized culture interacts vigorously with the theological heritage . . . "
>
> Walter Ong

As Tillich asserts, this language is not limited to spoken or written words, but it includes symbol and image. God's Word is more than words. To consciously or unconsciously limit Word to word is to weaken our relationship with God. Tillich declares, " . . . the Word of God often is understood – half-literally, half-symbolically – as a spoken word, and a 'theology of the Word' is presented which is a theology of the spoken word. This intellectualization of revelation runs counter to the sense of the Logos Christianity. If Jesus as the

Christ is called the Logos, Logos points to a revelatory reality, not to revelatory words." [2]

In the same way visual arts have been identified with idolatry, so, too, can words, doctrines, and verbal constructions be idolatrous when they are elevated at the expense of the visual. This is what Tillich calls "the Protestant pitfall." To suggest using visuals in worship is a way to balance the mediation of God's revelation to our human senses. Just as God's revelation is not confined to image, neither is it confined to word.

Introducing visuals in worship is a starting point for reclaiming the fullness of the Logos as "revelatory reality."

Reclaiming the Power of Image in Worship

A typical worship service is saturated with words: hymns, printed prayers and responses, sermon, lyric and anthem. These claim our fullest attention. Even as the eye sees architecture, movement, banners, bread, wine, and water, singers, and speakers, these visuals are de-emphasized as words are emphasized. The ear has plenty provided for it, but the eye never seems to get enough.

One Sunday morning I illustrated a sermon with a video clip from a trip to France. In the clip were scenes of members of our church relating with members of another congregation in Paris, with whom we have a sister relationship. As I turned on the television set in order to show the video, I looked out at the congregation and noticed all eyes turned and focused on that blue screen. In that instant I saw eyes hungry for visual feeding.

Harvey Cox has advocated the importance of the visual arts in general and film media in particular as expressions of "the theological import of the visual world." Recognizing the Protestant emphasis of "the Word at the expense of the Light," he suggests that "it may be time to redress the balance a little." [3] Protestants have emphasized the creative power of word and de-emphasized the creative power of image.

When we de-emphasize visual imagery in our theological discourse, we miss that which precedes verbal construction: image. A word-oriented, book-oriented church misses the power of the image and its essential cultural and theological contribution.

To understand imagery and symbolism as media for God's revelation is to legitimately explore the potential that electronic visual arts of film and video have for deepening our experience with God's revelatory power. These arguments suggest the importance of including within the life of the church, and specifically within its worship life, visual arts, including electronic visual media.

The church's mission is to communicate the Gospel – to make disciples – in every generation. To do that, the Gospel must be communicated using the available technologies of each generation. Twenty-first century preachers have additional means for communicating power messages with the advent of electronic and digital communication systems.

Incorporating audiovisual media in worship engages our senses, encourages our attention, and deepens our response to God's work in our lives. Walter Brueggemann suggests this is critical to our work: " . . . the task is to fund – to provide the pieces, materials, and resources out of which a new world can be imagined . . . people in fact change by the offer of new models, images, and pictures of how the pieces of life fit together – models, images, and pictures that characteristically have the particularity of narrative to carry them. Transformation is the slow, steady process of inviting each other into a counter story about God, world, neighbor, and self." [4]

Film and video engage the eye, the ear, the heart, and the mind of the viewer. They transport us to different settings – and encourage us to make choices about how we live our lives in response to God.

Theology as Theophany

This discussion about reversing our bias towards word in favor of a fresh look at image leads to an examination of our theological language.

The very term we use to talk about our experience with God, "theology," reflects a bias towards "word," theos-logos = "God-words." Other words loaned from the Greek reflect our bias, too: We value logic (*logos*) over fancy (*phanos*).

The favored status of our word "theology" can trace its roots to Greek Stoic philosophy and early church tradition, which refer to a threefold way of speaking about gods in natural, civil, and ritual functions. It wasn't until Abelard in the 12th century that *theologia* was used to refer to "a philosophical treatment of the doctrines of the Christian religion . . . ."

The term "theology" came to express our systematic thinking about God. With the development of the printing press, rational, linear, word-oriented contemplations found a medium for distribution, and the printed word became a means for the spread of the Protestant revolution in the 16th century. The revolution was fueled not only by a suspicion of the abuses of the Church, which visual arts and architecture represented to many, but also by a technology that allowed for mass distribution of words.

God's "Word" came to be mediated through words and logical constructions of such words. Theology came to be understood to mean "The study or science which treats of God, His nature and attributes, and His relations with man and the universe."

Tillich concludes his theology of revelation with the striking declaration that all of the "different meanings of the term 'Word' are all united in one meaning, namely 'God manifest' . . . ." While he doesn't use the Greek word, he is talking about Epiphany, which means "manifestation," or, literally, "to show upon." Using biblical Greek to translate Tillich's phrase, "God manifest," we arrive at theos-phanos, or "to show God."

The difference between theology ("the study or science" of God) and theophany ("a manifestation or appearance" of God) is the difference between transcendence and immanence – a study removes us one step from the experience, while a manifestation is direct experience. The very word "theology" is more transcendent, more removed from us, than the word "theophany," which is more immanent, more immediate.

To propose using visual arts in worship is to reclaim the fully immanent revelatory power implied in the term "theophany," which is defined as "a manifestation or appearance of God or a god to man [sic]." Just as the church uses words to understand God's revelation, so does the church legitimately use pictures and imagery.

Through the use of visual arts and electronic visual media, the Word of God is mediated in fuller expression. Using visuals in worship is a way to rebalance our need for word – and our need for image – as mediators of God's revelation to humanity.

Theology is theophany, theophany is theology. Word and light are joined to one another in an interactive dynamic. Biblical narrative weaves this theme throughout its stories.

The interplay of light and word begins in Genesis 1:3, "Then God said, 'Let there be light . . . .'" Light is called into being by the creative word of God. Light emerges in relationship with God, with darkness, with heaven and earth. Similarly, the Gospel of John pairs word and light, "And the word became flesh and lived among us, and we have seen his glory . . . full of grace and truth." (John 1:14)

At the burning bush, Moses is engaged both by light (the fiery bush) and God's voice, the word. In the great stories of the exodus from Egypt are many dramatic examples of God acting through light and sound. At Sinai, there are lightning, thunder, the blast of a trumpet, and a voice. (Exod. 19:18-19)

At the baptism and transfiguration of Jesus, the voice of God is paired with light playing off of the figures of a dove at the baptism and of the prophets Moses and Elijah during the transfiguration.

The interplay of light and word is evident in Paul's conversion. On the way to Damascus, a bright light and a questioning voice encourage Saul's conversion: "Now as he was going along and approaching Damascus, suddenly a light from heaven flashed around him. He fell to the ground and heard a voice . . . ." (Acts 9:3-4)

Later Paul writes, "For it is the God who said, 'Let light shine out of darkness,' who has shone in our hearts to give the light of the knowledge of the glory of God in the face of Jesus Christ." (2 Cor. 4:6)

To bring together Word and Light in worship by using electronic visual resources, paired again as they have been since the dawn of creation and throughout important biblical stories, can effect a richer evangelism and teaching ministry.

Theology that includes multisensory phenomena opens a deeper and wider experience of God's revelatory Word, a Word that is known in word and images. Worship that engages spoken and written word and still and moving visual imagery assists in our honoring and glorifying God.

In the next chapter we'll discuss communication theory and a way to develop congregations expecting something to happen in worship through the concept, already touched upon in Chapter 1, on how to build interpretive communities.

Footnotes

[1] Background on the oracle and process is provided by Rollo May in *The Courage to Create*, New York: Bantam Books, 1976), pp.111ff.

[2] See Paul Tillich, *Systematic Theology* (Chicago: The University of Chicago Press, 1967), p. 157.

[3] See Harvey Cox, *The Seduction of the Spirit* (New York: Simon and Schuster, 1973), p. 268.

[4] See Walter Brueggemann, *Texts Under Negotiation* (Minneapolis: Fortress Press, 1993), p. 20.

Chapter 3

Building Interpretive Communities

The church is all about communication. The first verses of the Bible tell the story of God's communication with the primordial chaos: "the earth was a formless void, and darkness covered the face of the deep . . . then God said, 'Let there be light,' and there was light." In those early verses from the book of Genesis (1:2-3) we find a communicator (God), a message ("Let there be light."), an audience ("the face of the deep"), and a response ("and there was light . . .").

Jesus teaches his disciples and the gathering crowds with parables, with stories. He communicates a message to an audience and asks response to it. The early churches – and every subsequent generation of Christians – have used various means of communication to tell others the stories of faith and to invite people into a relationship with the love of God in Jesus Christ. The stories, doctrines, traditions, and teachings were shared through oral communication, written letters and sermons, manuscripts "illuminated" with hand-painted illustrations of the text on the page, mass-produced Bibles and pamphlets, visual arts of sculpture, painting, fresco, mosaic, etching, the musical arts of lyric, melody, and harmony, and later photography, motion pictures, and animated stories.

Modern communication theory tries to analyze the various parts of communication, primarily message-sending (the message itself, the sender or communicator, the means of communicating [oral, print, electronic]) and message-receiving (the audience, its understanding of the sent message, and the audience response to the message).

Drawing on the work of Paul Soukup, S.J., in his article, "Understanding Audience Understanding," I want to help us become aware of some of the assumptions people within our congregations may be making about the communication process when screens and visuals are introduced into worship.

The abstract of Soukup's article defines our territory:

"On the one hand, the 'powerful message' construct paints the audience as passive recipients of the meaning presented in the media. On the other hand, the 'active audience' construct places most interpretive power in the audience . . . . A middle position sees audience understanding emerge from an interaction between messages and audience members."

Powerful Message Theory

This view of media was developed in the 1920s as mass media were growing in power and it was understood the audience was merely a passive recipient of these persuasive and effective messages. The assumption was that "if the message was sent and the audience exposed to it, it would have the intended effect." Audiences were understood to be passive and, in some ways, powerless under the effect of a strong "bombardment" of messages. If the communicator carefully constructs the message, it was thought the audience would automatically receive the message and understand it entirely.

"The new generation, raised on TV and the personal computer but deprived of a solid primary education, has become unmoored from the mother ship of culture."

Camille Paglia

There is certainly some truth to the idea there are "powerful messages" that have an engaging, gripping, attention-holding quality to them. Many have concerns about sophisticated, immersive, professional-quality messages that create powerful responses in audiences, particularly unsophisticated audiences such as children or those who choose to be ignorant or divisive in their views of life. Powerful media can display violent, militaristic, racist, consumerist, and sexist programming in a way that is embraced, imitated, and accepted by audiences. These messages and those who communicate them are part of the content of life, and intended and unintended audiences will want to be prepared to address them. This is an aspect of the "Powerful Audience Theory" of communication.

Powerful Audience Theory

The other pole of this way of understanding the communication process sees the audience as having the most power, and this was based on a view that the audience selects the messages it will receive – and will create the meaning of the programs "based on their experience rather than on the presented meaning of the media source." This view of the audience began in the 1970s and continues to this day with the increasing popularity of audience-generated programming on Web sites such as YouTube and the decreasing market share of the ABC/NBC/CBS television networks.

Powerful audiences have the remote in their hands, ready to change channels frequently and quickly if the programming or message doesn't suit audiences' interests or needs. Intelligent audiences know they have the freedom to expose themselves or not to various communicators and their messages. Audiences know they have "psychic shields" that have sensitized audiences to messages they wish not to experience (such as pictures of starving people, blatantly sexist lyrics in a song, or violence in any form) and can "turn off" the messenger and the message with a change of channel or a mental shield. The rise and fall of various programs and personalities shows how audience tastes and interests change quickly, always forcing programmers to adjust their messages and means of communication.

Messages in Worship and Preaching

Traditional congregations accustomed to speaking-listening styles of worship with lots of music and an oration for a sermon will be concerned about the power of the message only as far as how well it is presented. A "good sermon" will be perceived as a powerfully delivered piece of communication from the preacher, and a "boring sermon" will be perceived as having less power. The congregational "audience" wants a powerful message. Interestingly, congregations that are good listeners never understand themselves to be "passive" because they know how they choose to listen carefully to interesting and well-presented messages, and such congregations also know how they "tune out" the preacher when things get slow or uninteresting. How many times have people confessed to planning out their week while "listening," or to thinking about the brunch menu, or visualizing their golf putt up the center aisle! In the oral presentation dominant worship, there is an interactive balance between the powerful message and the powerful audience.

However, when screens and visuals are introduced into worship, something different happens. A few critics will be quick to see that screen as no different from a television screen or a movie screen and give to that church screen the same sort of negative evaluation they give to TV and the movies. The critics will associate the church screen with the "powerful message" theory they have adopted in regard to the mass media they have seen and heard and for some reason will not see their generous exclusion of oral presentation from their critical framework.

For these critics it is perfectly acceptable for the congregation to be considered an "audience" but not to be "spectators."

Assumptions about the congregation as audience members:

They sit and listen.

They pay attention.

They concentrate on what is being said.

They go inside themselves and process what they are hearing.

They sometimes respond to music they hear with applause.

Major distraction: listening is interrupted by noise from children, loud sounds such as coughing, traffic outside, or other noises from heat system, etc.

Spoken communication invites intellectual growth.

Sermons are "meaty" and provide intellectual food the mature and educated can digest.

Listeners are active rather than passive.

Assumptions about the congregation as spectators:

They sit and watch.

They are being entertained.

They are passive onlookers.

Visuals are the "spoon-feeding" of coddled or powerless children. Adults can listen, children look.

Listening goes more deeply than looking.

Major distraction: looking is interrupted by other visuals such as architecture, the human speakers, and other movement in the congregation.

Visuals invite shallow emotional response.

A Third Way: The Interpretive Community

In his article, Soukup suggests the two poles of communication audience research are extreme positions that help us understand the forces that act upon messages and people. A middle or third way of understanding takes into consideration the fact that messages and audiences are always in relationship and affect and change each other. As Soukup writes:

"Meaning results from the actions of both. Programs/texts do carry meaning, meanings which their creators did in fact intend. Audiences for their part do actively negotiate meaning, based on, for example, their positioning, their prior experience, and their needs. Communities of interpretation offer another means to understand audience understand of programs/texts."

Messages and audiences interact, whether they are in oral communication settings of speaking and listening or in visual communication settings including screens and looking.

Worshipping congregations are, at best, communities of interpretation. Good listeners have always known this, but they may not have always understood their prejudice against looking and seeing.

Assumptions About the Congregation
As An Interpretive Community

It organizes around scripture, sacred music, liturgy, and praxis.

It engages the "kingdom" stories of faith.

It honors book and ritual.

It prays and keeps silence together.

It studies, discusses, and imagines, constructing alternative visions of life through faith.

It experiences the arts together: paintings, sculpture, music, film, dance, fabric arts.

It grows in trusting one another.

It develops means for whole-brain learning.

It formulates ways to engage the world through faithful witness.

Principles of Interpretive Communities

In Chapter 1 I shared the unexpected results of using media in worship with a particular community of faith over a long period and the principles I learned as a result. Let me summarize them again:

Interpretive communities center themselves around the scriptures. Interpretive communities grow as they center themselves in Bible reading to become acquainted with biblical texts. Beyond the reading of texts, the communities find ways to study the Bible together to learn about the ancient contexts out of which the texts grew. The communities find ways to engage various art forms – literature, art, film, and music – to discover ways that scripture and culture interpret and inform each other.

Interpretive communities are grown over time. Those who join with a worship community participate in that particular community's history, tradition, and practice. This is a living process over a period of time. People have come and gone in the life of that community, and the community is always changing through births, marriages, confirmations, and deaths as well as through its participation in its entire social, educational, mission, and worship experience. People grow comfortable in these communities by spending time in them.

Interpretive communities equip one another to grow in understanding. Interpretive communities learn to present and interpret the messages of Christian faith through the educational, social, worship, and mission practices of the faith community to which the communities belong. They not only continue to experience the world around them and the communities in which the interpretive communities live, but they also find ways to select social projects that fulfill the interpretive communities mission as a church. As these communities engage in the practice of ministry, they also gather in worship to continue to build their community of faith, honor and worship God, and send one another out into the world equipped with faith and purpose.

Conclusion

It is easy to fall into assumptions about a world of powerful messages that seem to render powerless their recipients or that audiences are able to effectively resist and control manipulation or propaganda. A better way might be to consider building "communities of interpretation" that actively engage, interpret, and construct the meanings that a broken world desperately needs.

In the next chapter we'll look at how people of all ages are accessing more and more information with a variety of screens and at how interpretive worship communities can develop cross-generationally.

Note

The Paul Soukup article, "Understanding Audience Understanding," is found in *From One Medium To Another: Basic Issues For Communicating the Scriptures in New Media* (Robert Hodgson and Paul A. Soukup, S.J., eds., Kansas City: Sheed and Ward, 1997).

Chapter 4

Worship That Engages All Generations

In the previous chapter we talked about how the church's central mission is communication, mirroring the very first verses in Genesis and the story of God's interaction with the deep chaos. God as communicator spoke a message in the form of a command to bring forth light, and something heard it ("the face of the deep"?) and responded ("and there was light . . . ").

Later, in Exodus 3, God catches Moses' eye with Divine Fire at the burning bush. As Moses comes closer, the Holy Voice engages him in a discussion. God then sends Moses on a mission to tell Pharaoh to "let my people go." Jesus taught his disciples and the gathered crowds with parables and stories. He communicated a message to the people in his audience and asked response of them.

The early churches, and every subsequent generation of Christians, used various communication methods to tell others the stories of faith and to invite those people into a relationship with the love of God in Jesus Christ. The stories, doctrines, traditions, and teachings were shared through speaking and preaching; written letters and sermons; manuscripts illuminated with hand-painted illustrations; mass-produced Bibles and pamphlets; visual arts of sculpture, painting, fresco, mosaic, etching, and photography; musical arts of lyric, melody, and harmony; and later, motion pictures and video vignettes.

The church's mission is to communicate the Gospel and to make disciples in every generation. To do that, teachers and preachers communicate the Gospel using the available technologies of the current generation.

One hundred years ago, Christian clergy in the United States promoted the use of emerging motion picture technology in the church to communicate the Gospel to children, youth, and adults in worship and religious education as well as to promote mission awareness and involvement. While some church leaders shared this interest in the new technology, there were others who raised the caution flag, seeing danger in the "via media." Those who advocated using the new moving-picture technology as a tool of the church countered by pointing out church people have always opposed technical changes before adopting them, including such things as quartet singing, the pipe organ, the printing press, and new translations of the Bible.

The situation hasn't changed much when we speak of worship and technology today: There are many who embrace it, and there are others who oppose it. The very word "technology" has become a barrier, a metaphor for a kind of rapid change that isn't always good or helpful. Yet experience shows the technology people allow into the sanctuary is the technology that serves the worship functions of the church.

Churches have already adopted many different technologies into worship space that have become virtually invisible because they have been so well integrated into the worship space and experience. These include the following:

Architectural Technologies

These provide structural building features such as foundations, walls, and roofs, along with windows and ceilings; lighting fixtures; heating and cooling systems; furnishings such as altars, tables, pews, and chairs; and chair lifts and elevators. Included in this list might be the decorative technologies placed into sanctuaries such as those that facilitate fabric art and banners, plant and floral arrangements, pictures, artwork, and seasonal displays of altar arrangements.

Liturgical Support Technologies

These include sound systems with amplifiers, speakers, and microphones; organs, pianos and other musical instruments; clocks (we all know how the clock has quietly changed expectations about service length!); communion elements (bacteriological studies influenced a shift from wine to grape juice); printed worship bulletins, songbooks, and hymnals; audio and video tape-recording systems and broadcast capabilities; and so on.

Presentation Technologies

While these might be understood to be liturgical support technologies, I think they deserve special attention. They include projectors of all kinds (slide, filmstrip, 16mm, LCD video, etc.); surfaces for image projection, including walls, screens, fabrics, etc.; computers and presentation software such as PowerPoint; flat-screen televisions in place of projector and screen; and data sources such as 35mm slides, videotape (analog) or DVD movie, computer (digital) generated text and visuals or even Internet brought into the sanctuary with wireless signals.

Recent surveys show the prevalence of presentation technologies in today's churches. In an April 2008 report, the Barna Group cited a study on church use of emerging technologies, including worship technologies of large screens showing video imagery and movie clips.[1] The study found 65 percent of Protestant churches have a large-screen projection system. The study also found the rate of adoption of these systems has slowed since 2005 and the large-screen system is related to a church's size and theology. Among churches with less than 100 adults in weekly worship, only 53 percent use these systems, while 76 percent use them in churches with 100-250 adults, with upwards to 88 percent of churches drawing more than 250 adults to worship each week.

The survey also found only 43 percent of churches described by the pastor as having "liberal theology" have the big-screen possibilities, compared to 68 percent of churches self-identified as theologically conservative. We are left to draw our own conclusions on reasons for this until more data are gathered about why this is true. Eighty-eight percent of churches now with big screens in place show movie clips or other video segments, which is more than in 2000 but less than in 2005.

George Barna, the director of the studies, speculates that small churches may be less technologically friendly because of lack of size or budget and they just might be small because of that kind of self-limiting thinking. In addition, he says that as more digitally resistant churches find ways to fit their vision to the use of these technical tools, there may be further growth in the use of such technologies.

This expected growth in churches using projection technologies is seen in the recently released Duke Divinity School Congregations Report (October 2014) in which congregations with projection equipment tripled between 1998 and 2012 from 11.9 percent of congregations to 35.3 percent.

Because of the wide use of presentation technologies across the United States, most church members have experienced worship with the technologies in place. If people's own churches don't employ technologies, people have seen technologies used when people have visited other churches for special occasions such as baptisms and holiday celebrations with family or just out of curiosity to see what other churches are doing. My experience is what these visiting church members report when they get home is less an observation about technology and more about what they experienced: how they liked singing with their heads looking up at a screen rather than down at a hymnal; how they liked not having to hold a heavy hymnal; how they appreciated not having to follow along with a worship bulletin in an unfamiliar church and could just follow along on the screen; how the pictures and video clips they watched connected with the mission and message of that church; and how they noticed a lively spirit among those who had gathered for worship.

Another recent study shows that just as more and more churchgoers are becoming accustomed to presentation technologies in worship, so too are people increasing the personal use of various media technologies in daily life. The Nielsen Company, known for its data on television viewing, issued its first quarterly "Three Screen Report" regarding television, Internet, and mobile usage in the United States in May 2008.[2] What Nielsen found was that those aged 55 and older watched an average of 168 hours of television a month, twice as much as children and teens and 36 percent more than 18-54 year olds.

As a percentage of the video-watching audience (defined as video seen on live television or playback), the 55-plus age group accounted for twice as many viewing as children and teens. Those 35 and older accounted for 63 percent of the video viewing audience in comparison with 37 percent of those under the age of 34.

Now called the "Nielsen Cross Platform Report," the 2014 quarterly reports show a steady increase in the total of hours people of all ages spend watching content on a variety of screens.

When we put these data side by side with Barna Group studies on church attendance, we find the age groups watching the most television and video are also more likely to be attending church. A

2006 Barna Group study showed 54 percent of elders (born before 1945) and 49 percent of boomers (born between 1946 and 1964) attended church services, while 43 percent of busters (born 1965-83) and only 33 percent of mosaics (born 19842002) attended.[3]

Where we might think our older members would be against the use of presentation technologies in worship, and media in general, the older people are in fact the very ones who use media technologies at home more than other age groups.

Stated in other terms, someone who turns 100 years old this year was being born as churches in Connecticut, Indiana, and California were advocating showing movie scenes in worship. Documents from 1909-1916 show how clergy and laity from several denominations were promoting the use of motion pictures in worship and education.[4]

Today's 91-year-old was born in 1924. That was the year my grandparents were married, and for their honeymoon they went to see a silent film in Chicago. Today's average 91-year-old (and there are more and more in today's churches!) was born as churches in the United States had been using motion pictures in education and worship for just over 16 years.

While the Nielsen studies show people across the generations are comfortable with television, movies, the Internet, online video, and mobile media, there are still discomfort and resistance about using presentation technologies in worship. There are many reasons for this, including the resistance of a lot of tired clergy who are overextended in their ministries already and can't see adding something new to what they're already doing and including church members who regularly see how poorly presentation technologies are used in their workplaces. One need only go to www.youtube.com to view short videos featuring how not to do PowerPoint presentations!

If all these people who come to church already are familiar and comfortable with television and other media use, why is there such resistance in some congregations, particularly self-identified liberal congregations, to the use of such technologies in worship? I am convinced what we've done is trained clergy and developed congregations that have come to expect worship is for our ears and not our eyes. We've developed congregations of auditory learners who expect that worship is more about listening and hearing than it is about looking and seeing.

Consider how most Protestant worship has been mediated by the aural arts of music and preaching. This 500-year-old emphasis on speaking and listening has pushed aside the worshipful use of visual arts of painting, sculpture, film, and photography.

The reasons for ear-centered worship go back to Exodus 20:4 and its cautioning against graven images and extend through the Protestant Reformation with its elevation of the spoken, preached, and printed word as the best means to produce faith. As a result, most Protestant worship today is designed for the ear. Most of what happens in worship – music and spoken words – treats the ear. We take it for granted our ears are the instruments of faith, and we give them music and preaching that are pleasurable, instructive, and worshipful.

> "As we become a more visual culture, children and later generations of adults will respond more fully to what they see than to what they hear. Or better, they will have difficulty hearing what they do not also see. Protestant churches which ignore this fact will probably become increasingly marginal."
>
> John Westerhoff

The ongoing National Congregations Study through the Hartford Institute for Religion Research (and now through the Duke Divinity School) has reported in the United States 90 percent of a typical worship service involves listening to choirs, liturgists, and preachers.[5] Worship has become a listening culture where people come to hear a good sermon and listen to good music.

While statistics vary, studies show that of today's adults, 70 percent are visual learners, 25 percent learn by hearing, and five percent are hands-on learners. Is this true of our worshipping communities? Given the viewing habits shown by the recent Nielsen report, this is probably so. Given the resistance many pastors report to attempts to use video clips and other visuals in worship, it might be our church practice has created congregations full of auditory learners – or at least of people who have grown accustomed to an ear-centered worship culture.

Age alone does not determine which sense is dominant, as many youth and young adults are also auditory learners. I remember a faithful member of a congregation I served who closed his eyes during the entire sermon. It wasn't that he wasn't listening, since he could tell me specifically what I had said. This auditory learner was simply blocking out all visual stimuli so he could concentrate on what he was hearing. I imagine that were he alive today, he would oppose any use of film or photography in worship on the grounds it would be distracting to his concentration. During a particular confirmation class, I recall a couple of teenagers who, instead of watching a class video, sat with their heads on the table, eyes closed, a picture of lazy inattention. When they correctly answered every question I asked, they proved that despite appearances they were listening very carefully to the video. They were auditory learners and did not need to see the screen to know what was going on.

Our churches are full of such auditory learners, and since aural content has been the predominant practice of many churches, most worship communities are composed of people who prefer an auditory approach to worship. Many of these church members would find it distracting to look at pictures of any kind. These are the people who, when hearing a suggestion to show a video clip during a sermon, would say, "If I wanted to watch a movie, I'd go to the movie theater."

That said, many churches have successfully integrated more visuals into historically audio-centered services by projecting visual announcements, words to hymns, images that enhance the sermon, and classical religious art. There is evidence that congregations, even auditory ones, enjoy the pairing of sound with sight when done tastefully, sparingly, gracefully, and gradually.

These experiences are further strengthened when presentation technologies are used to celebrate the intergenerational nature of the church. For example, some churches have easily integrated into their services pictures and video clips that show:

– pictures of the couple that was just married in the church the previous weekend

– photographs of the latest newborn with mom and dad, grandpa and grandma

– pictures of the whole family gathering after a baptism

– scanned images of art projects from a church school class

– pictures of the confirmation class as infants and now as they are at confirmation

– short video clips of a recent youth service project

– e-mailed photographs of church members on a mission trip

– screen shot of a community news item featuring a church member

– photographs of a recent senior citizen luncheon program

– photographs or video clips of church members who reside in assisted living facilities or nursing homes

These materials can be presented in worship during the announcements, as a special mission report, as part of the offertory, as an introduction to a special time of recognition, or during the sermon presentation. All ages of worshippers give special attention to these visual portrayals of the church family and friends.

Over many years of developing a visually rich worship experience, we heard many positive comments from people across the generations. One woman who visited with her children and grandchildren said, "I've gone to church my whole life, but this is the first time I got something out of it. It was because I could see what you were preaching about." Other long-time members who had attended worship weekly came to prefer the visual service because they could understand more of the message with its combining sound, words on a screen, pictures, religious art, and relevant movie clips. Attentiveness increases with younger members, too. Infants and children quiet when an interesting visual is presented. I recently saw a 2-year-old stand up on the pew and lean forward to see the pictures that were illustrating my sermon. She followed along intently.

There are other positive intergenerational results that can emerge from the convergence of worship and presentation technology, including increased attendance, attentiveness to worship content, and mission awareness and response.

A visually centered and culturally relevant worship experience creates an interest among those searching for fresh and meaningful worship. Children, youth, and adults start coming to church with the expectation something new, different, and engaging will be presented as part of the worship service. My experience with a visually rich service showed that young adults and families with children started attending worship more regularly. They represented a generation accustomed to screens, relevant music, and film clips, but for whom the application of these materials to biblical story and theological theme was exciting. Members invited friends and family into the worship service. Grandparents began to report things like, "I came to the media service because this was the service my children and grandchildren attended. I also came to like it myself."

Another response to the use of visuals in worship was increased discussion and interaction with material presented during "visual sermons." With the addition of visual illustrations, including symbols, diagrams, film clips, and photography, people could reflect upon their own experience with what they saw and heard and gain a memorable experience to talk about afterwards with others. Pictures made the preaching themes more memorable by adding an additional sense – the visual to the auditory.

Effective visuals help people better understand the mission projects the church promotes. Pictures of these projects, and in some cases brief video clips, help draw people into the project and stimulate financial support for these very specific needs. After Hurricane Katrina we raised thousands of dollars because we were able to show pictures of local people on site in Louisiana purchasing and delivering diapers, towels, and blankets. In another instance, hundreds of dollars were quickly donated to help a local family with hospital expenses associated with the premature birth of twins in a nearby city. We were able to show pictures of the newborns on a weekly basis and report their progress as well as highlight the amounts being donated by church members toward various medical needs. Parents reported more conversations with their children about what they had seen on screen, and there was a growing church family awareness of the opportunities for serving the needs of others.

Developing An Intergenerational Interpretive Community

Probably the most important change that took place as a result of the introduction of media arts in worship was the development of an intergenerational interpretive community. As more and more people became engaged in direct experiences with creating liturgical media art,[6] those persons came to understand the importance of having theological conversations to link biblical text with cultural media productions such as film clips and pairing of images to music and lyrics.

As youth and adults became comfortable developing PowerPoint presentations with original photography and appropriate music, we found teams of fathers and daughters, mothers and sons, daughters and mothers, and fathers and sons offering to provide pieces of liturgical media art.

Rather than becoming recipients of someone else's messages about scripture during worship, people became direct participants with a stake in a creative engagement and application of scripture to cultural media. People developed fluency with a kind of hybrid worship language from the creative interchange between cultural media and theology.

Introducing visual presentation technologies into congregations accustomed to an emphasis on auditory technology can be jarring. Yet with sensitivity to the audio and visual needs of congregation members of all ages, the introduction can be done gradually and well. When such changes are harnessed to the purposes of worship – to gather around, experience, and respond to the Word of God in words, music, imagery, sacrament, and offering – the change is understood as serving the heart and soul of the worship community. Worship becomes engaging for all the generations, and the technology that helps make this happen remains invisible.

In the next chapter we'll look at a history of screens and how churches have been using screens, in one form or another, for a long time.

Footnotes

[1] "New Research Describes Use of Technology in Churches," April 28, 2008, and available at http://www.barna.org

[2] This is from "Nielsen's Three Screens Report: Television, Internet, and Mobile Usage in the U.S." (May 2008). See http://www.nielsen.com for the latest quarter's report of the "Three Screens" survey results.

[3] See http://www.barna.org

[4] See Terry Lindvall, *The Silents of God: Selected Issues and Documents in Silent American Film and Religion, 1908-25* (Lanham, Maryland, and London: The Scarecrow Press, 2001).

[5] This is data from the 1998 research of the National Congregations Study through the Hartford Institute for Religion Research http://hirr.hartsem.edu/cong/research_ncs.html. Findings for the 2006-07 and 2012 surveys have recently been released as the Duke Divinity School Congregations Report. All the data by the institutions during the periods may be found on the Internet at

http://www.soc.duke.edu/natcong/Docs/Changing_American_Con
gs.pdf.

[6] For further discussion about the developing field of liturgical
media art, see Eileen Crowley's *Liturgical Art in A Media Age.*
(Collegeville, Minnesota: Liturgical Press, 2007).

Chapter 5

Screens In Worship:
"Framing" the Conversation

As worship communities move toward incorporating more visuals in worship, there are many questions about the projection screens that may be needed: how and when to use them, where to get them and in what size and material, whether it's even a good idea to use them at all, and even why not to use them!

While all of these questions are important ones to any congregation thinking about screens in worship, another conversation lurks behind it all: What is this thing called a "screen?" What does it have to do with God and the worship of God? Could it be the church has always used "screens" but we just haven't noticed?

Simply put, a projection screen is a framed surface upon which are projected colors, shapes, designs, pictures, and words.

The earliest surfaces used for expressing messages with color, shape, and picture were cave walls, where an available space was used by early artists to call attention to an essential event of daily life: the hunt. Early Christians used a similar surface, the walls of Roman-era catacombs, to portray essential events of daily life: eucharist and prayer.

The surfaces of these walls provided small spaces that were dedicated to the content – and the meaning – of human life. Some early wall art had no discernible frame, such as a free-formed picture of a bison, while later wall art, such as Michelangelo's Sistine Chapel ceiling, was framed by a combination of the architectural features of the ceiling and the lines of color the artist used to distinguish one panel of the frescoed mural from another.

Church interiors today employ a number of framed surfaces for communicating content and meaning: plain painted walls that sometimes include murals or pictures; colored-glass windows sometimes showing symbol, picture, and story; carved or painted wooden panels with story and symbol; stretched canvases painted with landscapes and human forms and framed as pictures; hanging fabrics with sewn or applied shapes, colors, symbols, and words.

Given this wide variety of framed panels already included in church interiors, the projection screen may be seen as another framed surface the church uses to display the content of religious faith through color, shape, pictures, and words.

A Surface for Storytelling

For millennia, people have understood the storytelling power of figures fashioned from light and shadow. Plato's "Allegory of the Cave," dating back 2,400 years, uses the analogy of a wall with shadows cast upon it by firelight to suggest a role of education is to help people understand the difference between what is real and what is a representation of reality. Philosophers and storytellers alike came to see the educational and artistic possibilities brought by a light source, a surface, and the casting of shadows.

Walls and other surfaces would become the means, through light and shadow, by which skilled performers would bring delight – and moral lessons – to audiences throughout the world. These artists tapped into the interplay of light and shadow and found ways to use shadows themselves as storytelling devices for fun and learning.

For at least 2,000 years magicians and acting troupes traveled through China, India, Indonesia, Asia Minor, and Europe sharing the craft of shadow puppetry, bringing their audiences imaginative worlds full of entertaining stories and morality tales. First using walls and later fabric screens, some of these performers understood the screen to be God's universe upon which they cast shadows of their hands and handmade puppets as characters in a divine drama.

The English word "screen" finds its origin in this relationship of light and shadow. Hundreds of years old, the first uses of the word referred to upright panels covered with leather, cloth, or heavy paper and set in front of the hearth to form a room divider and shield people from the direct heat of the fire. It's easy to imagine children sitting in the space between the fire and the screen, using their hands to make shadows on the screen and telling grand stories about these figures. These screens, like those of the shadow puppet theaters, became a place for playful imagination.

Techniques for intensifying and focusing lighting effects developed gradually over time. Early light projectors were constructed to shine light through painted glass and to show the pictures on room walls. One of the first light projectors was called a "magic lantern," and it was used to delight small audiences with picture stories shown on walls, sheets, and special fabric or paper screens. Images were painted on glass in various colors and projected on the wall of a small room.

In 1646 the Jesuit priest Athanasius Kircher wrote a paper called "The Great Art of Light and Darkness" and gave instructions for building one of these light projectors. While he loved to "astonish" his viewers with this new visual art, he also encouraged them to understand the images were not magically produced but occurred naturally through the relationship of a light source (a candle or sunlight), a mirror inside the projector, and the wall or screen upon which the colored pictures were shown. Eager to connect his projector and screen to his theology, Kircher is said to have actually traced rabbinical use of projected images back to Solomon's Temple in Jerusalem.

"People want screens because screens open the door to more stories, more images, more information, and more excitement than ever before."

Kevin Roberts

As photography and motion picture technologies advanced in the 19th and early 20th centuries, churches began to use technically impressive projectors and screens to raise awareness and develop support for important mission projects. In 1908 the Foreign Christian Missions Society used a screen and projector in the Central Christian Church of Indianapolis to show church delegates stereopticon slides and picture films of missionary work in Japan, China, India, and Africa.

With the continued advancement of early motion picture technology, a number of clergy and laity in the United States advocated using movies and screens in worship services. Thomas Edison gave his blessing to these projects with an article he wrote for a church periodical in 1910.

During the course of the 20th century churches began to use other types of projection equipment in education and worship, including filmstrip projectors, 35mm slide projectors, opaque projectors, and transparency projectors, and projected picture and color onto walls and screens developed for that purpose.

A Window to the Holy Imagination

A screen, then, can be any surface upon which or through which light, shadow, and color may be projected. The screen itself may be a wall, a piece of colored or translucent fabric stretched on a wooden frame or attached to adjacent walls by taut lines, an unrolled synthetic white or gray surface stood on a tripod, or a specialized material suitable for video/data projectors raised and lowered from a narrow case by means of electric motor. The screen may consist of a reflective material upon which light is cast, similar to a typical filmstrip or slide-projector screen, or a translucent material such as thin fabric or a synthetic material best suited for video/data projectors. With translucent material, the light projector may be in front of or behind the screen.

The screen becomes a window for seeing the world, as the whole world can be shown on and through this window. It becomes a panel for displaying God's universe through the relationship of light and word, as at the beginning of creation when God said, "Let there be light." The screen becomes an artistic canvas for church artists to develop their holy imagination to show the worshipping community the relationship of God, Jesus Christ, and the Holy Spirit with all of creation.

Framing it in this way, we might begin to see a screen as just another framed surface on a church wall.

Think of the screen as the wall, the page, the panels, the window, the mosaic, the painting, the fresco, or the fabric art. The screen becomes any of the framed surfaces the church has used to show its story with pictures, symbols, and words:

A catacomb wall

A printed page of words

An illuminated manuscript with word and picture

A painted wood altarpiece

A stained glass window

A mosaic

A stretched and painted canvas

A frescoed wall panel

Fabric art

A framed photograph
A movie screen
A flat screen television
A screen can show clear and large any of these classical art forms:
A picture of praying figures from the catacombs
A page from any Bible or illuminated manuscript
The panels of a medieval wood altarpiece
A stained glass window from any church
A mosaic from the apse of a basilica
An oil painting of a bible story
A frescoed wall panel from the time of Renaissance
A piece of fabric art
Photography of important subjects

Looking at screens in this way, that what is shown on them fits the worship, educational, and mission life of the church, and a church can display a different content on its screens than that experienced by most people on the screens they experience at home, at work, and at school.

The apostle Paul wrote about the relationship of light, darkness, and the Christian community in his letter to the church at Corinth: "For it is the God who said, 'Let light shine out of darkness,' who has shone in our hearts to give the light of the knowledge of the glory of God in the face of Jesus Christ." (2 Cor. 4:6)

Light projected through and upon worship screens helps the community grow in its understanding of the content and meaning of Christian faith. The screen can become a means for the church to be in relationship with people who are eager to learn more and to grow in faith. Through engagement with the world and the positive influence of the creative arts to sensitize and grow awareness, people hear and respond to a calling to ministry in fresh, new and ever imaginative ways.

In the next chapter we'll look at common questions raised by clergy who just aren't convinced of the value of introducing screens and visual arts into worship in this second decade of the 21st century.

Part II

Feeding

St. Luke's United Methodist Church, Dubuque, Iowa

These next three chapters address common questions and concerns leaders and congregations might have in introducing more visual technology and art into their worship services.

Chapter 6

Common Clergy Concerns
About Using Digital Media in Worship

When clergy get together and start talking about the trouble they are having getting a congregation to change, they too often make the assumption that change is blocked by laity. My experience has been that too much change is blocked by the clergy themselves! After leading a number of workshops and seminars around the country, I have found that while clergy are interested in adding media to worship and changing worship practice to make it more relevant, exciting, and attractive, there is little interest in following through to make it happen.

There are many reasons for this: Clergy are too busy and already overcommitted with what they must do that adding another new program or innovation will stretch them too thin; some are too set in their ways and are unable to muster the energy to retrain to lead worship and preach differently; some are afraid they will have to change the way they do things; some know they don't work well in teams but are better equipped to develop worship alone; and some are afraid they will lose theological control.

Here are some questions, statements, and assumptions clergy commonly have about using media materials in worship, along with some suggested responses for helping relieve the concerns:

Question: How much time will it take to get a media service started?

Response: You can begin next week, or you can take your time and carefully work it through the life of your congregation. Here is what happened at my church in a nutshell:

a. For five years I gradually introduced visuals into our worship experience by projecting 2-3 slides during a sermon to show the biblical geography to which I was referring, by adding a photograph to a bulletin insert and calling attention to the photo in the sermon, and by using "The Art of Christmas" filmstrip during a sermon to illustrate different aspects of the Christmas story. Doing this required neither special permission from committees nor money. It's always helpful, however, to alert your worship committee members to what you are doing and why and to hear their responses once you've tried something.

b. I did receive permission to provide a six-week Saturday afternoon service that used a video clip, a work of art, and a CD song to illustrate my message. This took no money, and since it was an experiment, there was nothing "permanent" about it to threaten those who don't like change.

c. It took another two and a half years before we began a stand-alone multimedia worship in our church.

d. That said, another option is to begin now and not delay. Start using visuals in worship to grow your own comfort – and your congregation's comfort.

Moral of the story: Take all the time you need.

To ponder: Reflect on the above timeline with your own church in mind. Develop a sequence of steps toward implementation and a projected timeline based upon your congregation's (and minister's) readiness and adaptability to change. Identify the resources you will need and the people who may be able to provide assistance with the various steps in the chart below.

Question: Since our church committee structure moves slowly, do you need a committee to approve your using an occasional visual in a sermon?

Response: It's always a good idea to consult with your worship group members and to tell them you want to make your sermons more interesting by adding some visuals. You could ask if there is any objection to showing a few pictures of the Sea of Galilee the next time you preach about Jesus' Sermon on the Mount. After you do that, ask for the committee members' response. Did it help the sermon or hinder? Ask if there is any objection to showing one piece of religious art during a coming sermon. For example, use

Rembrandt's "Prodigal Son" when you preach about the parable. Or tell them when you preach your next Communion sermon, you'd like to show a picture of Leonardo da Vinci's "Last Supper" and maybe compare it with the Last Supper painted by Caravaggio or even Salvador Dali. You're not asking for an "act of congress" here, just bringing the committee along slowly to experience the power that visuals add to our worship and learning.

To ponder: Identify the committees that need to be involved in discussions about making changes to your worship service offerings. Make note of whether this involvement needs to be informative only or whether permission needs to be obtained through this body. Track the process in the chart below.

Statement: Technology is a bad word for some people in our church.

Response: Don't use the word! The issue isn't technology, anyway! The issue is communicating the Gospel. Here's a story for you. When I told my members at a local nursing home I would bring my computer and projector to show the members pictures of what was going on at church, one woman said, "I won't come." I asked her why. She said, "Because of computers." She did not want to be in the same room with a computer. Shortly after the conversation with her, I brought the computer and projector to the group and showed pictures of my trip to sites Paul had visited in Greece and Turkey. The woman loved the program. I never mentioned there was a computer in the room. To her, computers represented something bad. What she saw was an old-fashioned slide show. The screen was the same screen the church had used for decades to show filmstrips and slide shows. What she didn't notice was the slides weren't in a carousel projector but were on a computer hard drive. The projector connected to the computer used a lens and a bright light, just like the old slide projectors. She never opposed projectors and slide shows. She just doesn't like computers and couldn't articulate what she opposed.

It is important to note the church is not in the business of promoting "technology." What we are doing is communicating the gospel and using what used to be called "visual aids" and "audiovisual resources" as we always have done. It is your responsibility as the leader of the church to express the vision as clearly as you are able. You must identify your vision and your purpose. It must be rooted in the integrity of prayer, scripture, and theological reflection. Using media in worship is not about technology but about the Great Commission. It's about evangelism, revelation, incarnation, creation, and discipleship. It is not ancillary ministry but firmly connected to the core of faith.

Statement: It takes money, and we don't have any!

Response: No it doesn't. You have available technology in your church closets and in your members' homes. Use what you already have to introduce visuals to worship. Ask your members how you can improve upon it. Borrow equipment, or ask them to use their equipment at church. Rent equipment for a demonstration. Plant the seeds for the equipment you will need – identify it and its cost. Put the plea out there. Donors love to provide this stuff. At the present rate of change, they also have used equipment that's perfectly acceptable for our use at church. You could have five computers in a jiffy by just asking for donations of equipment in people's basements. Get software at institutional costs. Ask members to install memory chips into your computers. Get your people involved, not your committees, especially if your committees are designed more for talking and less for accomplishing.

Question: We're a small church and we don't have people around who can do this.

Response: Two places where you'll find people to help out are in the public school systems and your public library systems. Both places have people familiar with computers and other digital technologies. Several churches I know have father-son teams working, with the father coming out of the school system's AV department and the son knowing all about sound systems, computers, software, and where to find good music. There are also people connected to our congregations who are inactive because the church just doesn't speak their language. Do you know how many families in your church have digital cameras or digital video cameras, use the Internet, watch movies, and/or know the latest music? Do you have any art teachers or artists in the congregation? Many of our visual learners (see the section on auditory and visual learners) just haven't been asked to share their gifts with the congregation. If you still can't find people in your church, there will be someone in the community with whom you could form a relationship to help you get started! Pray about it, and see whom God sends!

To ponder: Identify individuals in your church who may be able to offer assistance with skills listed above. Make your list as extensive as possible, even if you don't know whether they will be willing to assist with the project. Ask other congregation members to name individuals who may have an interest or expertise in each area. For each person you identify, ask him or her to offer the name of 1-2 other people who may be able to offer assistance in one category or another.

Question: How much time did it take you to do that? (Hidden statement: If you say anything more than one hour, I don't have the time for this!)

Response: Adding media to worship is a team sport. No one individual can do this alone. Jesus defined a team: "wherever two or three gather in my name . . . " A team can be as few as two and as many as . . . twelve (the other number Jesus gave us!). Or more! By sharing the responsibilities and tasks, visual arts can be introduced to worship. There are times when there isn't time or the team just can't come together or help out, and then you do what you always do: Use your oral skills to communicate the message. Just don't give up on a vision of regular and frequent use of visuals during worship and preaching. Another part of this is that clergy need to evaluate their use of time. Clergy know how their time is wasted or used inefficiently on time wasters such as extra meetings and non-worship responsibilities. Use the 80-20 rule: Put 80 percent of your time on the thing that is only 20 percent or less of your week – worship. Worship is when you see the most people in the church; when you have the most impact on the most people; when the community gathers to celebrate, give thanks, and recommit to God; and it is the time when you receive your greatest revenue. If you are putting a majority of your time to worship

> "If a few people could only stop asking whether this is 'a good thing or a bad thing' and spend some time in studying what is really happening, there might be some possibility of achieving relevance."
>
> Marshall McLuhan

planning, preparation, development, and presentation, how will all the other things you do get done? Who said the clergy had to do everything? Find ways to involve the congregation in those tasks: Train small groups of people and delegate to them some of your responsibilities. Let the church do the work of Jesus Christ, and let the clergy focus on one of the primary responsibilities the congregation has called them to do: prepare and lead worship. Sure, this is easier written and said than it is done. All things are possible, right?

To ponder: How do you usually answer the question of "how much time is involved?" for the things you need to do? If you're aware the time commitment might be more than you would care to handle, how might you be able to reduce the time commitment involved? Who else might be able to assist with this task?

Statement: Only city churches can do media worship. (Hidden assumptions: because "city people" are smarter? have more money? understand technology? are more educated? are more inclined to support the arts? are less conservative?)

Response: Examine your assumptions! Anyone can do this, not just "city churches." My experience beginning, developing, and sustaining a multimedia worship ministry was in a church in a small conservative town in rural Wisconsin (population 6,000) where the average educational level was high school or less and where most people had a long experience with traditional church and worship. This is why I say, "If we could do it, you can do it." We did not budget for multimedia ministries for six years, and yet the funds came in whenever we needed anything because the people caught the vision of what could be done. Whether a church is in the city or the country, what church people don't want is someone coming in and applying some model for change on them that disrespects their history, tradition, mission, and purpose. As I transitioned a mid-sized rural congregation to accept and welcome the use of visuals in worship, I didn't use a model for it as much as I found models that worked. My single "model" for navigating change in a congregation grew out of my personal experience in ministry: that congregations can and do resist change and that people and churches like to be alerted to a coming change and be given the reasons for the change while being consulted along the way. This comforts and assures those who need assurance the change is an organic part of their life and seeks to serve the church's purpose with a minimum of disruption. Explanation and repeating the explanation go a long way. So do having a rationale and expressing it frequently, connected as it should be with scripture, the work of the Holy Spirit, and the history and tradition of the congregation. Church leaders and church members respond best when they are consulted, asked their response, given direct experience with some of the changes being discussed, and asked of their opinion. Responsive leaders will make any necessary adjustments along the way. That's the long answer. The short answer is: Work with laity to make change happen, broaden the ownership of the change, and it becomes a part of the fabric of a congregation's life, no matter if it's a city church or a country church, a large church or a small one!

Statement: If we let the lay people help with this, the clergy will lose theological control.

Response: This question shows some clergy think they have been educated and trained to the extent they know more than the laity in their churches – and the clergy's more sophisticated theological experience is the "right" interpretation. Clergy fear that if laity can select imagery to put on the screen or decide what popular music could be used in worship, the material would be theologically inappropriate or misleading. Other clergy subscribe to Luther's view in the "priesthood of all believers." The way out of this is continual dialogue and an ongoing effort to become a "community of interpretation" where everyone has access to scripture, tradition, theological doctrine, ecclesiastical agendas, and congregational experience and shares together in the work of making meaning. The statement shows an orthodoxy or "right thinking" that must be distinguished from "wrong thinking." The answer says we work together, clergy and laity, and provide the "check and balance" of a theological system that invites all believers, clergy and lay, into full communion and community.

To ponder: What statements in your church's covenant, creed, doctrine, purpose, and/or mission statement support the practice of including lay leadership in providing your worship services? Document the evidence contained in your church's background documents.

Clergy aren't the only ones in our churches who have serious reservations about using screens and technology in worship. In the next chapter we'll look at resistances among other members of the congregation and what to do about these concerns.

Chapter 7

Identifying and Overcoming Resistances to the Church's Use of Technology

When introducing worship technologies into church sanctuaries, leaders will want to be aware of the questions about such technologies and be prepared to respond to the objections. What follows is a summary of the criticisms and their responses.

There aren't many congregations left in the United States that worship in caves by candlelight, if there ever were any. Our churches have employed every technological innovation used in the other settings of our lives: electric lights, central heating and air conditioning, padded pews, electronic sound systems, electrically powered organs and pianos, elevators and chair lifts, fans dropped from ceilings, speaker systems attached to walls and ceilings, stained glass windows, energy-efficient building materials, duplicating and copy machines for worship bulletins, mass-produced hymn and prayer books, tape-recording systems, compact disc hymnal accompaniment services, and the latest building designs.

Technologies that provide centralized heat and air conditioning on demand are unquestioned, even as precious natural resources are consumed and often wasted in increasing amounts by worship communities.

Where once worshippers customarily stood to worship, bench seating provided a new comfort, and a culture used to soft easy chairs at home now demands cushions in pews.

Electricity-requiring worship aids such as lighting, sound systems, pipe organs, printing equipment, computers, typewriters, and clocks all play a role in religious sanctuaries.

Similarly, wireless microphones give freedom to clergy and laity who wish to be heard by those with hearing deficits while speaking from various locations in the sanctuary. Soloists who do not have time to rehearse with pianists and organists use recording equipment for prerecorded accompaniments; small churches without piano or organ, or persons to play them, use prerecorded accompaniments for hymn singing and other music.

Technology to make non-fermented juice from grapes allowed churches to respond to a growing prohibition movement working to keep alcohol away from its members. Innovation in technology allowed for the replacement of wine by grape juice, introducing a new communion practice that replaced 1,900 years of tradition.

The several aforementioned examples of the already widespread adoption of various types of technology in the church sanctuary led to a consideration of video technology in the sanctuary as a natural extension of other technologies already in place.

Typical Objections to Screens and Video in the Sanctuary

1. The very presence of a screen is a reminder of the numerous forms of consumerism, commoditization, sexual stereotypes, banal entertainment, and violent action that emanate from the television screen 24 hours of each day.

2. Using screens and video resources in the sanctuary encourages the shortened attention spans in worship that television and the Internet have already produced at home and in school.

3. Using a screen and projecting a video image in worship invites passivity and non-participatory disengagement on the part of the viewer/voyeur.

4. As children and youth are exposed to increasing numbers of video resources in school, the church risks its own irrelevancy by using similar instructional techniques.

5. Introducing such resources produces the expectation they will continue to be used, developing a kind of screen and video dependency in the worship setting.

6. Worship is one place where people have traditionally engaged in the orality of speaking and hearing, of reading and responding, of honoring the tradition of hearing the word.

7. The Protestant tradition carries an iconoclastic suspicion of all images – sculpted, painted, photographed, televised, or projected onto a screen.

8. While there may be a new generation of electronically hungry parishioners, there are also several generations of print-oriented,

preaching-oriented parishioners who expect the proclamation of the word from the pulpit.

Responding to the Objections

These objections can be answered by those who take a less pessimistic view of the use of video technology in worship.

1. While using screens and projected video in worship can be a reminder of all that's wrong with such media in their commercial setting, using them in worship carries a responsibility for purposeful, instructional use. The worship setting itself naturally prohibits showing a television program or commercials – unless these contribute in some fashion, and in a limited time frame, to a specific worship message.

2. Using video resources in the sanctuary does not itself encourage shortened attention spans. If you think about it, most worship in Protestant churches is already a series of different events divided into short time segments. Worship services are divided into distinctive segments such as a call to worship, an invocation, an opening hymn, a children's sermon, a choral anthem, an offering time, another hymn, reading of scriptures, a sermon, a prayer period, a closing hymn, and a benediction. There is a rhythm of standing and sitting, speaking and listening, singing and hearing, sound and silence. Over the course of sixty minutes of worship each of these segments might take three minutes each, with a longer sermon. My point is that worship already breaks time into discrete segments and to introduce video into the worship context recognizes video would be absorbed into this format.

> "I do not think that the powerful forces imposed on us by electricity have been considered at all by theologians and liturgists."
>
> Marshall McLuhan

3. While turning on a television set or projecting a video image can invite passivity over longer periods, when video technology is used in worship, it is used in short time segments of 3-4 minutes. These short periods protect against losing attention or encouraging some sort of passivity. Engagement with video would require verbal introduction and explanation by a leader, which would result in greater interest.

4. Parents and teachers report children and youth are exposed to video resources in learning settings, and it is true some parents and schools simply sit the children in front of a program and ask them to watch, without interacting with them to find out their reactions and learning from the program. Better use of video resources encourages dialogue with resources and stopping programs occasionally for stepping back to take stock of what is being said and learned.

5. While it is possible that introducing exciting video resources into worship can produce an expectation video will continue to be used, the reality is churches already use a variety of resources in worship on a weekly basis. If the chimes choir plays one week, do we expect the group to play every week? If other musical instruments or dance or a certain style of sermon presentation are used, we do not automatically expect these styles to be used week after week. Video would be simply an option among many options for calling attention to the messages of a particular worship service.

6. It is important to recognize worship is a time where people have traditionally engaged in the orality of speaking and hearing, of reading and responding, of honoring the tradition of hearing the word and that . . .

7. . . . . the Protestant tradition carries an iconoclastic suspicion of all images, but to fear the introduction of visuals in worship will minimize or eliminate the oral tradition of lyric, text, and print is mistaken. Visual are an enhancement that do not replace spoken language. Even as iconoclasm is a part of the Protestant tradition, most churches already contain visual arts, including stained glass windows, symbols such as crosses, fabric banners, paraments, and robes for choirs and clergy, altar arrangements, architectural configurations with symbolic meaning, etc. Using video opens a discussion on how a congregation has already demonstrated its commitment to the visual arts by looking at the visual symbols present in any worship service.

8. While there may be a new generation of electronically hungry parishioners, there are also several generations of print-oriented, preaching-oriented parishioners who expect the proclamation of the word from the pulpit, and this is a good reason to balance word with image, since worship settings most likely will include these different generations with different orientations in a worship setting. Just as the printing press did not eliminate speaking and storytelling, just as

television did not eliminate radio, so will computers not eliminate books. Preaching that uses a combination of resources such as illustrative material in worship bulletins, visual aids, and sermon outlines will find that pictures and moving pictures can also serve the message of the day.

It is clear our churches have been engaged with a dialogue between technology and theology for some time. To argue against a particular technology, such as using screens and projected visual imagery in the sanctuary, because of the potential evils of that technology is to ignore how churches have historically employed technologies in the service of God's mission.

Using the metaphor of how to get people to try new and different foods, the next chapter takes a look at how to incrementally overcome resistance to the use of visuals in worship.

Chapter 8

Picky Eaters in the Pew: A Strategy for Changing the Worship Diet

"Taste and see that the Lord is good."

Psalm 34:8

"I fed you with milk, not solid food,
for you were not ready for it . . . "

I Corinthians 3:2

Worship is a matter of taste ("aesthetics"), and by providing tasty foods we can also, as the Psalmist writes, see the goodness of God. Taste and seeing go together! So do aesthetics, the visual arts, and God! The Apostle Paul also understood that church people grow and change and that there is a natural "food progression" as we grow from spiritual infancy into spiritual adulthood. Milk serves our needs in our infancy, but as we develop and grow we need something more solid, and, tastier!

Let's play with these metaphors and learn how to develop the "visual taste" of our congregations through the gradual introduction of media arts in worship (a pedagogy of theological aesthetics?).

When trying to get a child to eat something new or even unpleasant to the child's taste, a frustrated parent might simply say, "Just eat it. It's good for you." Most of the time the child may not try it, unless the adult develops a positive strategy. Frustrated parents can turn to any number of child nutrition Web sites for guidance, including the Mayo Clinic (" . . . the dinner table can become a battleground . . . ") and "Getting Past Yech," a *Wall Street Journal* article about the picky eater.

It occurs to me that when it comes to worship, the worship leader and/or pastor is just like that frustrated parent trying to get a child to try something new: "Just give this worship change a try. It's good for you."

We understand that, like the dinner table, worship can become a battleground, to borrow the Mayo Clinic phrase. This being true, it doesn't mean we can't try to add something to the worship "diet" and even help parishioners grow in their appreciation for the new "food" of a worship change.

So, borrowing from the "Getting Past Yech" article, let's see how we might apply the same strategies to getting a child to try – and eventually enjoy – a different food to the similar task of encouraging a worship congregation to change its diet and try something new. Let's have some fun with this . . . and use your own imagination!

"When introducing a new food, put a small amount on the plate. Do this at least 10 times before giving up."

Worship media suggestion: When introducing media arts into worship, do so in small portions, and do it repeatedly. Be careful not to use too much of the screen at first. Show announcements before worship one week. Use pictures to illustrate a choir anthem the next. Show a very short mission message from a denominational videotape or DVD during the offering. Display three or four "Last Supper" paintings during communion. What you are doing is giving the congregation small amounts of media arts without a "full meal" and just enough to grow everyone's level of comfort and familiarity.

"Let the picky eater wash, cook, handle, or even play with new foods. The idea is to familiarize the person with different looks and smells and to reduce fear of the unknown."

Worship media suggestion: Notice the family members who take photographs after a wedding or baptism of a member of your church family. Ask if those persons could e-mail the church a few of those photographs for display on the church screen. Display the photos as part of the announcements within the week or month. Add text to the photo that congratulates the family and identifies the newlyweds or the newly baptized as part of the church family. You will be connecting the family to the congregation, affirming the use of digital photography by the family, helping it and the whole congregation grow accustomed to the use of a screen in the sanctuary. One message is that the screen isn't used for something that is "done to them" from the outside but is a communication medium from within the congregation. No longer is this something from "the unknown" or strange, but a part of the church family communication system.

"Try food chaining: Identify foods a fussy eater does like, then introduce other times and build from there. If the finicky eater loves chicken nuggets, try breaded nuggets of other meats, and then move on to meat that's not breaded. If vegetables are the issue, start with crispy vegetable chips, then move on to baked sweet-potato slices."

Worship media suggestion: Having developed a sense that what is displayed on-screen is imagery from the church families, add a little more. Show pictures of a confirmation retreat or a recent women's group meeting or congregation members working at a food pantry or at another community-based activity. Use these pictures during the visual announcements before the worship, and slowly add them into places during the worship such as the offertory period or perhaps as an illustration of mission during the sermon presentation. You are using local imagery and pictures of the life of the congregation during the message times of the worship service.

> "The great moments of civilization . . . come when cultures are open to other cultures, when they recognize that it is only by sharing that they can grow, grow richer in experience, and become something more than they were."
>
> Kim Veltman

"Cover new foods with a familiar sauce that the reluctant eater already likes."

Worship media suggestion: Here is an opportunity to use visual imagery while the choir sings. The anthem is a "familiar sauce" that adds to the "flavor" of worship. By selecting imagery that fits the lyrics of a choir anthem, you are enhancing the flavor of the music, amplifying (in a visual way) the choral contribution to worship by adding visual imagery, and visually offering the congregation members another way to experience what they are accustomed to. Alternatively, a preacher might preach the sermon (something the worship community already "likes") with pictures that individually and visually "anchor" each point of the sermon. The congregation is used to a sermon and a preacher's typical sermon structure; the visuals, when selected well, increase attentiveness, add to understanding, and facilitate retention (remembering what was said).

"Don't yell or punish the picky, but don't cave in and cook them whatever they like either. Allow a fussy eater to go hungry occasionally to learn that his or her pickiness has consequences."

Worship media suggestion: Once you've started to use visuals, and once you start to hear some negative criticisms, don't give up! While your goal is to help the congregation become more familiar with this "strange new food" of visual arts in worship, after trying "ten times" (see the first suggestion!), you might want to stop for a week or more. People will notice, and some will begin to miss the visual "diet" you have introduced. They will tell you and begin to make comments such as "Are you going to use the screen again?" or "I miss the screen. You can use it every once in a while, you know." This "feedback" will help you understand you've begun to change the "diet" of the congregation members in a good way and they are beginning to enjoy the new flavors while appreciating the nutrition you are offering.

Food advisers tell us it's far easier to train a child to try a new food and like it than it is for an adult who has been set in his or her eating habits for a number of years. Maybe this is what Jesus understood when he said, "Bring the little children to me, for to such belongs the Kingdom of Heaven." Children, youth, and young adults will appreciate your introducing new visual "food" into worship, and, if our experience with worship change is an indicator, the older adults who have trouble with "new foods" will also grow in their taste.

Chapter 9

Copyright Law and Church Use
of Media Arts Resources

This section is meant to help worship leaders recognize the legal and ethical requirements of U.S. copyright law and to help you demonstrate an awareness of the intricacies of copyright law, while understanding the unique application to educational and worship settings.

Churches, like everyone else, must comply with U.S. copyright law. While not a "roadblock" to the use of multimedia in worship, copyright law is a "speed bump" that forces us to slow down and think about how we will legally and ethically use copyrighted material.

Copyright law in the digital age is a subject of debate as well as legal scrutiny. The Internet has made it possible for people to share texts, images, video clips, whole films, and music – and has opened up many disputes about ownership of this material and what may be copied, when, to what extent, and at what cost.

It is easy to "Google" for imagery, music, and video, to download the material, and yet to have no idea your legal basis for doing so. When you choose to use material you have not created in a presentation you are creating, it is good to know your ethical and legal boundaries.

Summary of Copyright Law and Media Arts

1. U.S. copyright law governs all use of copyrighted material. Churches must comply with the law. Many Web sites offer the full text of Title 17, the Copyright Act of 1976, along with helpful discussions about various details of the law and how they may be applied. All churches and their leaders should be aware of how their church measures up to the law's requirements.

2. The House Judiciary Subcommittee on Courts, the Internet, and Intellectual Property has been meeting since May 2013 " . . . to begin a comprehensive review of the U.S. copyright law in the coming months." This was prompted in March 2013 when U.S. Copyright Office registrar Maria Pallante called for a comprehensive update to the Copyright Act of 1976 and the Digital Millennium Copyright Act (DMCA) of 1998. There have been 11 hearings between May 8, 2013, and July 24, 2014, including one in January 2014 on "The Scope of Fair Use." It seems there will be little progress until the next Congress convenes in 2015 and more rounds of hearings continue, although more pressing issues facing Congress may set the review process back a bit. It is my view that we'll see little change in the Fair Use Guidelines for Educational Multimedia, nor will there be any intervention by Congress to limit or expand how churches use media in the course of their protected worship services.

> "U.S. Copyright Law is a "speed bump" that forces us to slow down and think about how we will legally and ethically use copyrighted material."
>
> Michael Bausch

3. The law allows "certain performances and displays . . . in the course of services at a place of worship," but other church settings require proper permissions and/or licenses. Blanket licenses for church music and for showing motion pictures are advised for the broadest possible coverage under the law. ASCAP and BMI advise that their recordings may be played during worship. On its Web site, ASCAP advises churches that "Permission is not required for music played or sung as part of a worship service unless that service is transmitted beyond where it takes place (for example, a radio or television broadcast)." BMI has confirmed the same in private correspondence.

4. "Fair Use Guidelines for Educational Multimedia" provides voluntary guidance for the amount of such material that may be used, the amount of time material may be stored before specific permissions are required, and for proper crediting of sources of material.

The Worship Exemption

U.S. Code Section 110 provides specific limitations to the exclusive rights of copyright holders and exempts certain performances and displays. Section 110 (3) of the code applies directly to the use of worship multimedia. Under this section of the law, churches are granted an "exemption of certain performances and displays . . . in the course of services at a place of worship . . . " This includes " . . . display of a work in the course of services at a place of worship or other religious assembly . . . " This worship exemption is a "narrow exemption" and must be carefully applied.

While the exemption allows for the use of certain material during worship, the provision does not apply to any other aspect of church life, such as educational (e.g., church school or youth groups) and social occasions. For example, if you wish to show an entire movie to a youth group or other educational or social gathering in the church, you need to have a license (see license information below).

Fair Use Guidelines for Educational Multimedia

In 1996, the Subcommittee on Courts and Intellectual Property, Committee on the Judiciary, U.S. House of Representatives consulted with hundreds of publishers, software companies, professional associations, organizations, libraries, governmental agencies, and educational institutions about the meaning of U.S. copyright code for educators wishing to use protected material in classrooms.

This consortium of interested parties agreed upon certain "Fair Use Guidelines for Educational Multimedia," whose purpose was to "provide more specific guidelines that educators could follow and be reasonably sure that they would not be in violation of the copyright law."

A final report was issued in November 1998 to offer "guidance on the application of the fair use exemption by educators, scholars and students in creating multimedia projects that include portions of copyrighted works, for their use in noncommercial educational activities, without having to seek the permission of copyright owners."

While the guidelines are not legally binding, they "represent an agreed-upon interpretation of the fair use provisions of the Copyright Act by the overwhelming majority of institutions and organizations affected by educational multimedia."

As you have begun to realize, while churches are not specifically mentioned under these "Fair Use Guidelines for Educational Multimedia," these voluntary guidelines can provide churches with guidance for their use of multimedia in worship. These guidelines are the only guidance we'll have, as I don't think Congress is soon going to be passing a law to give churches specific guidance on their use of media arts in worship!

The guidelines specify the portion of a work that may be included in a multimedia project. The rule of thumb is about 10 percent of a given work, whether it is print, movie, video, music video, or photographic images.

For example, the guidelines provide for using:

• Up to 10 percent or 1,000 words of text

• Up to 10 percent or three minutes of movies or video

• 10 percent but not more than 30 seconds of music, lyrics, and music video (However, American Society of Composers, Authors, and Publishers [ASCAP] and Broadcast Music Incorporated [BMI] say that churches may play entire songs "as part of a worship service".)

• Not more than 10 percent or 15 images from a single source.

What to Do About Various Media

Music: LicenSing provides a license to use copyright-cleared music for worship. Go to http://www.licensingonline.org.

Christian Copyright Licensing International (CCLI) is another music-licensing company, whose Web site is at http://www.ccli.com.

If you are using music from popular recording artists from formats such as CD or MP3, ASCAP and BMI affirm your use of such material "in the course of services at a place of worship." Check out ASCAP's Web site for licensing information and the organization's view of the church worship exemption to copyright law: http://www.ascap.com/licensing/licensingfaq.html.

Movies: Licenses for motion picture/video/streaming and music are available from CCLI, http://www.cvli.org/cvli/index.cfm, or the Motion Picture Licensing Corporation, www.mplc.com.

The license you buy is based on the size of your church and usually costs around $200 a year. The license allows you to show feature-length commercial videos produced and distributed by certain companies for use under certain conditions. These licenses apply beyond the worship setting and cover use in other social and educational contexts.

The Fair Use Guidelines provide that up to 10 percent or three minutes of movies or video may be used (without a license), meaning that clips should be three minutes or less in length. You can easily comply with this, since our experience is that clips of 90 seconds or less are often much more attention-getting and focused.

Video: Sites such as Google's YouTube.com offer millions of uploaded videos. This site is continually doing the work of assuring copyright compliance and clearance of such material, although the work is a constant challenge. For example, Torrent Freak reported on January 5, 2015, that copyright holders filed Digital Millennium Copyright Act takedown notices "to remove more than 345,000,000 allegedly infringing links from its search engine in 2014," a 75 percent increase over 2013. [1]

When using material from Internet-based sites, citing your source is a good idea, while using your own judgment about material that may be copyright protected, such as music videos, television shows, and movies. More and more film studios, television production companies, and recording artists are providing their material freely and without restriction, so be on the alert for these sources of legally available material. Clarifications, restrictions, and clearances seem to be released on a daily basis, so pay attention to digital media news!

Imagery: Exercise caution when downloading material from the Internet. The Fair Use Guidelines tell us that "Access to works on the Internet does not automatically mean that these can be reproduced and reused without permission or royalty payment and, furthermore, some copyrighted works may have been posted to the Internet without authorization of the copyright holder."

When using imagery from Internet sources, be sure to see if the Web site offers the imagery for free or if there are restrictions. Avoid restricted material, or be prepared to ask permission and/or pay a fee for use of this imagery. The Fair Use Guidelines provide the free use of five photographic/illustration images or not more than 10 percent or 15 images from a single source.

Other Suggestions

Develop Your Own Library of Visual Material – One way to avoid having to deal with others' material is to develop your own library of visual arts. Accumulate digital photographs and label and organize them around theological theme and scriptural reference. Scan 35mm slides you already own into digital files and store them along with other photographs. Be on the lookout for original artworks from artists you know, and ask permission to use their material in future presentations. Carry a camera with you at all times, and ask others to do the same, so you can add to your database of images. Learn how to use photo enhancement features so you can turn some of your digital photos into colorful and abstract works of art. Be ready to turn some of your color images into grayscale or black and white to experiment with mood and other effects. Adding to your personal and church visual library will provide you with a storehouse of material you legally own and control. For helpful guidance on legal aspects of photography, go to http://photosecrets.com/do-i-need-permission and read the article "Do I Need Permission?" from copyright attorney Dianne Brinson.

Include This on A Slide or in Your Worship Bulletin – An announcement declaring your awareness and observance of U.S. copyright law needs to be included in your program. A good place to put a PowerPoint slide with this information is in the announcements before the worship or at the end of your presentation. Printing it in any worship bulletin is also a good idea for those who may have missed the material during the slide presentation.

This language, or something similar, should be sufficient (as long as you've actually done this!): "Certain materials used in this worship service are provided in accordance with U.S. Copyright Act Section 110 (3) exemption 'in the course of services at a place of worship.' They have been prepared using the 'Fair Use Guidelines For Educational Multimedia' and are restricted from further use."

Learn More About Copyright Law – Go to major university library Web sites to learn about U.S. copyright law. For example, check out the Purdue University site at https://www.lib.purdue.edu/uco/, or go to the Stanford University site at http://fairuse.stanford.edu/.

The full text of the Fair Use Guidelines may also be searched online, with most universities offering the guidelines. For example, the University of Texas has the guidelines on its Web site, whose address is http://copyright.lib.utexas.edu/ccmcguid.html.

For a good print resource on churches and copyright law, get Richard Hammar's *The Church Guide to Copyright Law*. While Hammar does not specifically address the issue of media arts and worship, he does provide specific guidance about the church's legal obligations and rights pertaining to music, print resources, videotaping, broadcasting, and rebroadcasting.

An additional, more recently published resource is *Multimedia Law and Business Handbook* by J. Dianne Brinson and Mark F. Radcliffe.

In our next section, Part III, I offer a way to provide a visual style of preaching that can quickly engage a congregation. Each chapter in this section offers suggestions for thinking about what's in one's worship space and how this is connected to the nature and mission of the church.

Footnote

[1] See http://torrentfreak.com/google-asked-remove-345-million-pirate-links-2014-150105/

Part III

Preaching Architecture

St. John's United Church of Christ, Hartford, Wisconsin

The architecture of sacred space invites congregations to remember their connection to the stories and symbols of the faith of the congregations while reminding them of their central mission in the world.

Chapter 10

The Worship Space

" . . . tongues in trees, books in the running brooks, sermons in stones, good in everything . . . "

William Shakespeare

" . . . when your children ask in time to come, 'What do those stones mean to you?' then you shall tell them . . . so these stones shall be . . . a memorial forever."

Joshua 4:6-7

As Shakespeare poetically noted, the world around us is full of sermons waiting to be preached. Joshua told the people crossing over the Jordan River to put up a tower of stones to capture the children's attention. Seeing the stones, the children would ask why the stones were there, and the children's parents could then tell the story of the significance of that place.

Our worship spaces are similar monuments to memory, full of stories of the congregation and its reasons for being, as well as of the people who gave their time, energy, and donations to build and sustain the space.

Each sanctuary has stories to tell with mysterious symbols to describe, and, in a world where people ask few questions for fear of appearing ignorant, who better to ask those questions than a preacher?

A wonderful starting point for beginning a process of preaching architecture is to ask the children or youth what they notice in a sanctuary and to invite them to ask their questions about things the children or youth have seen. The young are curious, and, as visually oriented as they are, have noticed things that caught their attention.

A couple of children's messages inviting these questions about the sanctuary can then lead into a series of sermons about the architectural forms (layout, windows, furnishings, fabrics, symbols) and the stories behind the forms.

I've practiced this process with four congregations I have served. Each time I've received comments from people who say they have worshipped their whole lives in that sanctuary and never heard a sermon telling about the connections with the sanctuary and windows to biblical stories or have never heard anyone talk about the symbols the people have noticed on the sanctuary walls, woodwork, baptismal founts, pulpits, lecterns, and altars.

Worship leaders mistakenly presume people know the "language" of their worship spaces. What an opportunity it is to be able to help people look at what's right in front of them – and to see it fresh and new! The windows are telling stories, as is the altar or communion table. The position of the pulpit tells a story; the light fixtures just might contain hidden messages; and certainly where you have or have not placed the baptismal fount says something about the congregation's relationship with the sacrament.

Much may be gained from preaching architecture. It can:

• Capture the attention of those in the congregation who are curious, who have an interest and aptitude for buildings and architecture, and who have a personal investment in the building and upkeep of the space.

• Provide a valuable multisensory learning experience as the congregation focuses their visual attention on common elements of the church while listening to the preacher's observations and insights.

• Acknowledge those who have sat through worship services and focused their attention on some of the visual details in the sanctuary while listening to sermons, music, and prayers.

• Affirm the long history of a congregation as you speak of the origins of its church and the history of the building(s).

• Recall and affirm the positive memories long-term members have for their church.

• Develop a sense of unity and boost a sense of congregational identity as the congregation's relationship to the space is named and nurtured in new ways.

• Enrich the worship experience for those eager to learn more about the church they've chosen for worship.

- Heighten awareness of the particulars of a sacred space and stimulate reflection on what makes a space holy.

These spaces we inhabit for worship have been purposefully designed by church architects to serve the function of a worshipping community. Those who use the spaces give these enclosures different names. They might be called a "house of God" or a "temple," perhaps a "house of worship" or a "meeting house" or meeting place. The building might even be called a "church" or "chapel", a "synagogue" or "mosque."

The Sanctuary

For our purposes in this book, we'll use the term "sanctuary" to refer to a worship space.

The word "sanctuary" comes from the Latin "sanctus," or "holy." It is an enclosure of some kind that is considered a place set apart, a holy space within a part of nature or in a building. This space in nature may be bounded by a group of trees, or bushes, or stones, and in a building, by walls and a ceiling.

The term sanctuary speaks of a space that also may be subdivided into smaller, discrete spaces having additional functions within the larger space. Many sanctuaries include a narthex at the entrance to the worship space; a nave, where the congregation is seated or stands; and a chancel, where worship leaders gather around sacred furnishings such as an altar, communion table, or pulpit. Some sanctuaries also might include a smaller chapel area extending off to the sides of the nave.

Let's consider each of these spaces as found within a sanctuary.

The Narthex

The narthex, sometimes also called a foyer, is the entrance area between the front doors and the doors to the worship area or nave. The term "foyer" is French for the place of the hearth or fireplace, a focal point in a room. Some church foyers are very comfortable spaces with chairs, lamps, tables, literature racks, and a guest book. These all invite people into the "home" that is the church.

The Latin term "narthex" refers to a similar kind of area, with the additional meaning that once upon a time the area was a place to keep a smaller space separate from the nave reserved for penitents and others who were not welcomed into the nave. In a sense, it implies the entryway just outside the nave is a place of preparation for full participation in the life of the worship community.

The Nave

The nave is the central part of the worship sanctuary, where most people come to be seated, usually with an aisle or aisles separating two or more seating sections. The term comes from the Latin word for ship, which could refer to the story of Noah and the ark of salvation or to Gospel stories with Jesus and disciples in boats on Galilee.

Many sanctuaries look or feel like a ship, with the beams in the ceiling reminiscent of how one imagines a covered ship, such as Noah's ark. I've been in one sanctuary shaped like that ark that even had portholes for the entry door windows, and the porthole theme continued in each of the stained glass windows on the sides of the nave. Another sanctuary I know had blue-green colored windows to evoke the colors of the sea outside the "ark" of the church.

If a sanctuary evokes feelings of being in a boat, then some meanings we might draw would be how the humans who gather there are kept safe and nurtured for a while and then sent out again to the "dry land" of the world that lies fresh before them as they "disembark" as disciples.

While the building is of course anchored to the ground, the concept of church as a "boat" assumes a metaphorical kind of movement: The church is on the move rather than moored, and this boat navigates the waters of life while protecting its occupants from a remembered pre-creation watery chaos (see Genesis 1:1) or the deep of the abyss.

The Chancel

The chancel is the area around the altar, usually reserved for the clergy, choir, and lay leaders of worship. The term is from a Latin word referring to lattices or crossbars, a reference to communion railings or other wooden structures separating the altar area from the nave. Sometimes there are steps leading up, denoting going upward, and toward the sky/heaven or a sacred precinct. These steps may be a set of three to add to any trinitarian symbolism found elsewhere in a particular sanctuary.

The chancel typically contains an altar or communion table, along with a pulpit, a lectern, an organ/piano, and seating for clergy and the choir.

Spending Time With Your Worship Space

A worship leader wanting to preach the architecture of a holy space needs to spend time in that space to simply look around and pay attention to what is seen. Bring a notebook and pen or a recording device to record observations and maybe a camera to take pictures of what you notice.

A way to structure your experience might be to follow a "form-content-meaning" kind of process. Art historians sometimes use the terms "icon-iconography-iconology" to refer to a similar way to begin to analyze a work of art or architecture. The icon is the work of art itself, iconography is all that's written about that work, and iconology includes the meanings drawn from it all.

> "There is evidence that congregations, even auditory ones, enjoy the pairing of sound with sight when it is done tastefully, sparingly, gracefully, and gradually."
>
> Michael Bausch

Using this "form-content-meaning" method is a helpful way to begin to access all that is present in a worship space. Essentially the viewer identifies three things:

The *form* of the thing you looking at. Simply identify the kind of art or architecture it is (for example, painting, photography, fabric, sanctuary architecture) and the materials used to construct it. You are answering a basic question, "What is it? What is this thing?"

The *content* of the thing. What do you see? Look at it, and notice all the details of the work, such as colors, shapes, figures, and what you understand to be the story being told. The basic question you're asking is, "What's the story? What's going on, and why?"

The *meaning* of the thing and its story. This is where we start drawing out what the form and its content mean: what the form meant originally in the Christian story, what the form meant to those who included it in the sanctuary, what the form might have meant for those who worshipped here in the past, and what the form might mean for us now and down the road. What is the thing, the form, trying to tell you – and why? What does it have to do with God and the life, work, ministry, and mission of the church?

It is important that a preacher use all three aspects of this "form-content-meaning" process while experiencing the architecture, contents, and fabrics of the worship space. Sometimes the meanings are what come later, after further thought and the movement of the Spirit, and it is worth the wait.

To preach architecture, furnishings, and fabrics requires more than a simple description of what they are, because the listeners will want to connect with the stories that are told (the content of what you're looking at) and will welcome the meanings you draw as you move the listeners forward along their missional path as a worship community.

A Self-Guided Tour of A Worship Space

You might start by wandering the entire area of the sanctuary, starting from the main entrance, next into the foyer or narthex, and then to the nave and chancel.

Before you enter the nave, pause for a moment to remember this is someone's sacred space. When I've taken groups to St. Peter's Basilica in Rome, I've preceded them in and turned around so I could face them as they encounter the space for the first time: It is a moment of awe for them, and I see their eyes open wide and hear audible gasps. Other less grand worship spaces may evoke less of a response, and yet each sanctuary has its own spirit and testimony.

As you enter, experience the whole space for a while as you see it with fresh eyes. Take the time you need with this new introduction to your worship space and pay attention to your initial reactions with feelings and thoughts.

Consider what's there: the whole, and then the parts: the structure, shape, colors, smells, textures, flooring, windows, walls, ceiling, ornamentation, symbols, furnishings. As you're drawn into the space, notice the various "forms" of the things you see: What questions do they inspire, and what stories do they hold? What meanings might you draw from what you see?

After a few moments of this, remember you might be "preaching" your experience and what you see here, so make notes on themes and approaches you might use in any future sermon series.

After a walk-through from the entryway to the narthex and into the nave and chancel, you might choose to do any of the following. While this is a numbered list, you may certainly pick and choose these "exercises."

1. Sit where you usually sit and relax in place. Breathe, pray, meditate, and center yourself in that space. Begin to notice what you feel: your feet on the floor, the soft or hard surface of the seat, your back against the pew or chair.

2. Let your mind consider the history of the space, the years it's been used, the people who've been important to it. Think about the occasions that have brought people in: weekly worship, weddings, baptisms, funerals.

3. What energy is in the space, and how does it smell? Be aware of how all your senses are attuned to your experience there.

4. Look around. What do you see? Make a list of the items in the space and the questions you have about how the items fit into worship. What furnishings do you see? Ask yourself why these things are there?

5. What do you notice about the general order of the sanctuary: Is it clean, with everything in its proper place, or is it cluttered and full of things that don't seem to relate to each other in a meaningful way?

6. Look up to the ceiling. As you look at the structure of the ceiling, you might simply count the beams. Some churches will have three, evoking the trinity, but more likely six, evoking the six days of creation, while the whole structure itself makes the seventh, the day of rest and worship.

    a. Some churches have one kind of ceiling in the nave and a different ceiling for the chancel, or the area directly above the communion table, altar, pulpit, and lectern. While a beamed ceiling may rise above the nave, the ceiling above the chancel area may be plaster panels angled at each other, or even scalloped, to evoke a sense of sky or clouds while adding to the acoustical quality of the space.

    b. Sometimes the chancel area is enclosed in a large arch, as if to call forth the image of a shrine containing the "holy of holies," which might be the altar, the Bible, the communion table, and/or other furnishings that evoke the sense of that space set aside for the clergy, perhaps the choir, sometimes the organ and organ pipes, and usually a cross.

7. Notice the ceiling light fixtures. How many chandeliers are there, if any? One church I served had ten, with five on each side,

evoking the Ten Commandments and the light they shed upon a community of faith. When I pointed this out in a sermon in that church, the custodian told me there were four light fixtures in each one, making 40 lights. It seemed this further hinted at the Noah story, with the ark-like sanctuary and the blue-green sea windows, and the lights representing the 40 days and 40 nights. Without knowing if this were the original architectural intent, it did provide an interesting connection. Other sanctuaries have groups of six or seven fixtures, all playing off 6-7-8-12 symbolisms with the numbers (see No. 12 below for meanings of the numbers).

8. How many candles are in the sanctuary and where are they located? Why is this, do you think? Are they electric, wax, or paraffin? What are these – and their placement – saying about worship, liturgy, and light source?

9. Determine the orientation of the sanctuary. What direction does it face (north, south, east, or west)? What significance, if any, is this direction? Synagogues, for example, face Jerusalem. Some churches follow that convention. Still others are more functionally situated to take advantage of the morning or evening light with a south-north orientation.

10. What are the windows like? Are they clear, or are they colored? Do they contain stories, symbols, or color alone? Why, and what significance do they bring to the worship space? See the next chapter for more discussion on windows, colors, and stories.

11. Where does music come from? Where are any bells, chimes, organ pipes, pianos, guitars, drums, and sound systems located? How do these fit into the design of the sanctuary? Is there projection equipment such as a screen and projector, and how do these fit into the worship architecture? Are there any religious symbols connected to these items? For example, sometimes a central cross is affixed to the center front of organ pipes if they are located at the front wall of the chancel area.

12. Pay attention to groupings of similar things such as windowpanes, archways, or carvings in wooden backdrops or furnishings. Notice numbers of things: groupings of three might represent the Trinity, five might point to the Pentateuch or the first five books of Moses, six might represent the six days of the week, seven might represent the fullness of six days of creation and one day of rest, eight might represent a fresh beginning (the day after the

seventh day) or that which transcends both earthly reality (the number 6) and heavenly reality (the number 7), and 12 might represent the 12 disciples.

13. What symbols do you notice, such as Greek or Latin symbols or words on wooden or painted surfaces or fabrics? Common symbols are the Greek letters alpha and omega (the first and last letters of the Greek alphabet, representing beginning and end) and the chi and rho (the X and P, or the first two letters of the name "Christ" in Greek). You might also notice a cross with the Latin INRI, for Iesus Nazorenus Rex Iudaeorum, "Jesus the Nazarene, King of the Jews," as the Roman inscription ordered placed on the cross. One entire sermon could be devoted to these Greek and Latin symbols and their origins and stories.

14. A central symbol in Christianity is a cross. Is there a central cross and where is that located? What does the location seem to say about Christian faith, this particular worship sanctuary, and the gathered community? Is there an implied cross in the very shape of this sanctuary? Is the central aisle a long one leading to the chancel (evoking a Latin cross), or is it about the same length as the width of the sanctuary (a Greek cross)?

a. How many walls do you see? Sometimes what appear to be four walls is actually a structure with eight, representing the fresh beginning of the eighth day, the first day of the new creation, or the day after the Resurrection. Where is the center of the room, and what does that say about the architecture and the congregation that gathers there? Is the center of the room where the congregation sits, and what does that say about this space as "church."

15. Get up and move around, sitting in different parts of the sanctuary. Think of the people you know and where they sit, and go there to see what they see. Jot down what you notice about these different vantage points.

16. Are there memorial plaques showing names of various people over time who have given windows, bells, musical equipment, furnishings, founts, altars, hymnals, or other things found in the sanctuary? What stories are added to these features by the memories of these persons? What are some of the forgotten stories someone might be able to tell?

17. Make a note to return to the sanctuary after a worship service to pay attention to what your senses notice: Can you feel the warmth and pick up the scent of colognes and perfumes? What energy remains, if any? Consider the mystery of the gathering, what transpired, what it meant for all of you to have been there as a community of faith.

18. Come to the sanctuary at different times of the day and night to see what the different light or shadow adds or subtracts to the environment. What changes at different times?

19. Going deeper: What does the architecture reflect about the history of the congregation? Does the architecture fit the era when the church was founded? How does the architecture reflect the central Christian theological themes common to most churches? What interests and life stories of the founders are found in the sanctuary? How well do the structure and space fit the perceived mission of the church, and what do the structure and space say about the approach of the church to worship and preaching? Having old photographs of the space will help you grasp some of the history and passage of time.

Shapes in Church Architecture

Worship spaces come in different shapes. One shape might be a circle, representing the sphere of earth and the dome of heaven. This shape is found in mosques, often with a central chandelier coming down from a dome and resting just above the heads of the worshippers, representing the lights from the dome of heaven coming close to the congregation.

A shape with four equal sides is a square, which, like a circle, can represent an equality of balance and a reference to the four cardinal directions (north, south, east, and west). A space shaped as a circle, square, or octagon will have equidistant length and width, visible in a short central aisle. This may represent the equal arms of the shape of the Greek cross.

Sometimes what appears to be a square or rectangular shape has been altered with interior wall angles to become an octagon. This is very subtly done and sometimes takes a while to notice. A nave with an octagonal shape might represent a sense of equality as well as the eighth day of creation, a fresh beginning, when God and humans begin their renewed work together.

Besides noting the length of a center aisle, counting the number of interior walls in the nave and chancel might be a way to start noticing the theological shape to a worship sanctuary.

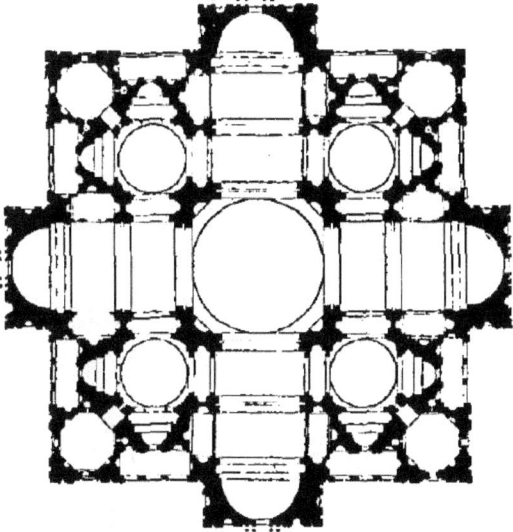

This diagram shows Bramante's Greek cross plan for St. Peter's Basilica, Rome.

A rectangular shape with four walls and a lengthy central aisle speaks more to the shape of a Latin cross. This hints at a formality and a focus on the leadership of the church, with the aisle used for ceremonial processionals and recessionals with clergy and choir.

The long rectangular sanctuary, as in a basilica such as St. Peter's in Rome, is shaped as a Latin cross, with the arms at the end of the nave where the chancel begins and the vertical of the cross extending from the entrance of the nave all the way to the front wall of the chancel. The long aisle facilitates formal processionals that end at the high altar, which is at the center of the crossing of the vertical and the horizontal arms of the cruciform shape. The high altar, where the Pope celebrates the mass, becomes the focus of the mass and the liturgy.

While these shapes may represent theological and liturgical commitments of a particular worship community, it is good to remember that the space itself is formed to follow particular functions.

Think about other public spaces such as lecture halls, concert halls, theaters-in-the-round, and amphitheaters, and then consider how your worship sanctuary compares. Notice how the form follows the function and how the space fits the human relationship of presenters and audience and the kind of engagement that will be invited. Will the audience become a part of the performance? Will the audience be expected to simply sit and listen? Will the audience need to hear well and see well? Is there a priority given to speaking and listening or to watching and seeing or both?

Where is the architectural center of the space, and what does that say about the purpose of the gathering and the design of what happens in that space? Is there a difference between the actual center of the space and the perceived center? That is, is the chancel the central focus and center of the space and event rather than the actual center somewhere in the audience?

In a worship space, what do the locations of these centers say about God, the community, Jesus Christ, worship, whom we are, whom we're called to be, where we're called to go?

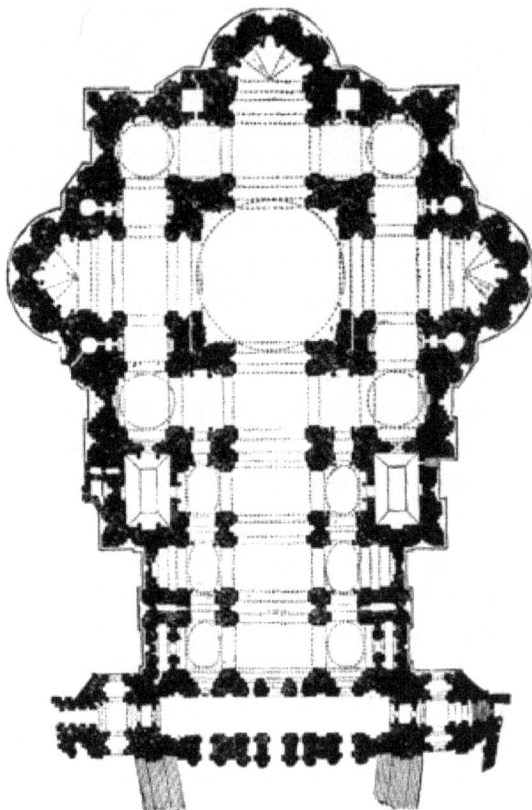

This diagram shows Rafael's Latin cross design for St. Peter's Basilica, Rome.

In the next chapter, we'll take an in-depth look at the chancel, furnishings, and fabrics of a worship space.

Note

Both images of the cross plans for St. Peter's Basilica are in the public domain.

Chapter 11

Chancel, Furnishings, and Fabrics

The Chancel

The chancel, as noted earlier, is the area to the front of the nave where the altar and/or communion table are located. In some churches the chancel is on a raised platform one or more steps up from the nave. Sometimes the pulpit and lectern are on the first level of the platform, with the altar/table another step or two above and behind. There may be a railing that separates pulpit/lectern from the altar/table, which makes that area by definition the chancel.

There are variations, of course, where some churches will position the communion table on the floor in front of the chancel or raised platform and still have an altar positioned at the back wall of the chancel. Where these furnishings are located is sometimes a mystery, depending on a congregation's customs, tastes, and personal decisions of various leaders.

Thinking about the difference between an altar and a communion table is important: The former is a place of sacrifice (from two Latin words "to make holy") where candles, a cross, and sometimes offering plates might be found. The communion table is where the eucharist is celebrated, where bread and wine are consecrated and shared. Sometimes there is only one altar/communion table with both functions served on the same piece of furniture.

The baptismal font is often placed in relationship to altar and table and is then a part of the chancel. In some churches the font is placed at the front entrance to the sanctuary, representing initiation into the church. In other churches, the font is located on the floor level below the platform/chancel to represent how the sacrament is a part of the community of faith, the gathered congregation. When the font is in the chancel, the font is connected to both word and sacrament, situated in relationship to pulpit, table, and altar.

Calling attention to the locations of these furnishings, their forms, and their functions – and their physical and theological relationships with each other – can be a sermon in itself.

Is there a central pulpit that serves as both the place where scripture is read and the word preached? Are there a lectern where the scripture is read and the liturgy led and a separate pulpit where the word is preached? On which side of the chancel is each located, and what does this say about that church's theology and worship? If there is a central pulpit and no lectern, might it indicate there is a unified and centralized provision of the word and its interpretation?

When there are a separate lectern and pulpit, there is a symbolic distance between both, and this might represent an interpretive freedom afforded the preacher: a space between from which the word is read and the word is preached. One might understand an invisible tension between the two with an implied difference between the Word of God (scripture) and the word of the preacher-interpreter.

There is also an interesting energy that extends from that space between pulpit and lectern out to the nave where the congregation sits, showing a relationship between scripture, preacher, and congregation together engaged in worship, liturgy, and preaching.

When preaching architecture, one might reflect on these energies as noticed in the furnishings, their function in worship, and their location in the chancel and, sometimes, the nave.

The Raised Platform

Shortly after the U.S. Civil War and into the early 1900s many of today's mainline churches (Presbyterian, Baptist, Methodist, and Congregational) used a Greek amphitheater style of architecture to bring the audience closer to the chancel and stage (platform) so the audience could better attend to the central event: preaching. Brought closer to the stage as with the ancient Greek theaters, people could better hear the preacher and choirs with a minimum of visual obstruction.

These protestant churches were beginning to also emphasize lay involvement in the life of the churches and were forming committees, structuring educational ministries, and building worship spaces that could gather this family of faith closer together in their common service.

Architectural books focusing on church construction in the 1880s sought ways to focus attention to the minister and the message while also offering the best sight lines, acoustics, and comfort for the congregation. Churches began to take on the look of classical concert and opera halls as music and preaching became central to worship life, and a kind of high urban culture of concert hall and church came together in urban areas such as New York City and Chicago and some rural areas as well.

A raised platform with three steps not only invoked the holy trinity and the elevation of holy presence but provided better acoustics in those days before microphones and amplifiers. At the same time, this helped meet the seeing and hearing needs of a congregation eager for worship, entertainment, and engagement.

Better acoustics and a sense of togetherness were accomplished with a wide platform and seating angled toward it. This also showed that worship is public and not just a private matter: Worship is best done together with other Christians. The space also provided a sense of awe and freedom with the upward reach of the ceiling and the ornamentation on the furnishings (even the clergy chairs had high backs with spires pointing upwards).

A Cruciform Shape of Nave and Chancel

As the Greek amphitheater became an architectural model for some of these free churches, they also adopted a floor plan closer to the pattern of a Greek cross, with its four equal arms. This pattern affirmed Greek themes that were matching up with the developing United States of the 19th century with its rhetoric of democracy and freedom as well as a new humanism of an equality before God. Theologically, the Greek cross with its balance, grace, and harmony could represent a common humanity before God and lift up the strength and contribution of the common man and woman gathering in the worship space.

In that shape we might see a symbolic image of God's power distributed among the people and shared with all rather than being concentrated in the hands of a few. While the clergy are leaders, there are delegation of authority and responsibility to the congregation.

Of course, not all churches were designed in the pattern of the Greek cross, as others, still having a platform and chancel, used the more elongated aisle forming a Latin cross. For these Protestant churches with a long central aisle, and those Catholic churches with a populist streak to them, the cruciform themes are still a part of the life of the church.

While the long aisle could serve grand processionals and highlight the grandeur and authority of the church, the Latin cross also is a reminder of the very nature of the cross: It was the Roman imperial instrument of torture and death for non-Roman rebels and seditionists, and it was upon such a cross (not a "Greek cross") that Jesus of Nazareth was crucified.

It is said we are to live a cruciform life. Jesus told us to take up our cross and to lose our lives for the sake of the gospel. To take up the Latin cross and remember the execution of Jesus by the powers and principalities of his day, we are called to remember all those who suffer and die for justice, righteousness, and peace in the world today.

To take up the Greek cross of equality and the heart of compassionate love is to lose our individualism and give it up for the sake of the community, to build a strong group, a strong church, a strong community.

> " . . . we believe . . . the design of the environment is a choreography of the familiar and the surprising, in which the familiar has a central role, and a major function of the surprising is to render the familiar afresh."
>
> Kent C. Bloomer and Charles W. Moore

The cross, Greek or Latin, is more than a symbol: The cross represents a form of thinking and living into which the church is called.

The church invited this theological theme into its life not only in the preached word but in the architecture of the space in which that word was preaching.

A Screen in the Chancel

The screen becomes a form through and upon which a community of faith conveys the content of the rich religious and theological imagination of the community. Part of the form, content, and meaning of the screen is its placement in the sanctuary – and the theological relationship of the screen to architectural elements already in place.

Where is the screen to be put in the sanctuary? Mostly, churches of traditional architecture find a place in the chancel area that is conducive to the temporary or permanent location of a screen. This might simply be a blank wall that serves as the screen surface, a portable screen that rests on a tripod or "legs" and is brought out for occasional use, a swath of fabric or cloth hanging above the altar, a framed screen that is permanently attached to a wall, or a motorized retractable screen that descends from a wall, a wooden beam, or metal armature.

The screen, in whatever form it takes, is always in relationship to what is already there, and that is generally a cross, an altar or communion table, and a pulpit.

The placement of a screen in relationship to cross and altar may be architecturally and theologically understood in this way:

• The altar or communion table is anchored upon the floor and the earth below, representing God's holy ground, God's protecting grace, and a community's thanksgiving to God.

• The cross is ever in relationship with the earth into which the cross is set and the heaven towards which the cross extends, and represents courage, suffering, strength, and love through the death, resurrection, and life-giving witness of Jesus the Christ. The cross symbolizes the church being called to a life of discipleship.

• The screen in relationship to cross and altar becomes the place of liminality, a doorway, a window, helping the community to imagine and live out its essential relationship to the divine story. Whereas the altar, communion table, and cross are fixed in appearance (although fabric draping, floral displays, and other temporary items may occasionally adorn the table and cross), what happens on the screen will vary. It becomes the place for showing words of prayers, scripture lessons, and hymn lyrics and for showing colors, shapes, symbols, pictures, and moving pictures. The screen shows stories freshly told and opens new ways for experiencing sermons, meditation, prayer, and instruction.

In some instances the screen might descend from above the pulpit. In this case, the screen becomes a part of the architecture of proclamation and is directly linked to the preaching and teaching function of the pastor. What is projected on the screen is understood to be part of the preached Word of God. The projected material supports and stimulates the interpretive imagination of both pastor and people by supplementing, illustrating, illuminating, and promoting the preached word.

Preaching the Chancel Furnishings

In addition to the altar, communion table, lectern, pulpit, and chairs, there are likely other furnishings and decorations in the chancel area and on the platform.

These might include:

Candles: note the number for any symbolic meaning and their location. Why candles? What do they evoke, and what symbolism do they recall?

How are they lit and extinguished? Who does this? When are they lit and when are they extinguished, and what does this say about worship in your space? Think again on the form (the thing itself), the content (What is the story? Why are they here?), and the meaning (how this connects with worship, with God, with the congregation).

These same questions of form-content-meaning would apply to the following elements:

Light fixtures

Flowers and plants

Projection screen

A cross

Organ pipes

Communion railings

How well do these fit together? Do the furnishings match? Is there a hodge-podge of wood colors and furniture styles? Does any of this add or detract to the worship experience? Are there ways to fashion more harmony and balance to what's on the platform? Is it enough to simply call attention to what's there in worship and preaching and to remind one another of the form, the function, and the meaning of what's there? Might these reminders also help a church decide what needs to remain and what might be placed elsewhere?

In a recent review of some of my photographs of the chancel areas of churches I'd visited, and in some cases served, I noticed how cluttered the chancel looked. In my direct experience with those sanctuaries I didn't find myself noticing the clutter, so why did I see it in the photographs? My answer, upon reflection, is that when we are present in a worship sanctuary we might in one moment presume an organic unity to all that is there, and in the next we focus on one object of attention. This focus keeps us from noticing the clutter, that which might be extraneous. One preaching strategy, then, could be to focus on the organic unity and then to focus on individual parts over time.

It would be interesting to bring in someone unfamiliar with worship spaces, hear the person's reactions to what is noticed, and draw lessons from the "unchurched" and whether one's space is inviting or off-putting.

It would be the preacher's choice how far to delve into these furnishings. Certainly references could be made along the way as a congregation begins to visually appreciate its worship sanctuary and learn to make connections with what is in the sanctuary with the worship of God and the nurture of the members' life together as a congregation.

Fabrics in Chancel and Nave

In this section I offer suggestions for other ways you can preach your worship space. A sermon or sermons could be devoted to highlighting the use of fabric in the worship space. Another set of messages could be devoted to the exterior of the church building and how its shape, color, and external features speak to what goes on in this place. Finally, you could also design a series of sermons on one or more of the artworks hanging throughout the church building.

Preaching the Fabrics

From time to time church members raise questions about the American flag and whether it is or is not to be displayed in the worship space. Many of us have served churches where the flags have been positioned in different locations over time, including not being in the sanctuary at all. With national attention often focused on flag etiquette, and who is or is not displaying flag symbols, these issues sometimes seep into church life.

At one church I served there was a question about the proper location of the U.S. flag and the Christian flag, whether they should be near the altar in the chancel or on the floor of the sanctuary, and on which sides, right or left?

Since we were approaching the July 4 holiday, it seemed like a good time to preach on the issue of flags in church, the history of their inclusion or exclusion in worship sanctuaries globally and in the United States, a variety of theological positions on the subject, and what this congregation's response might be.

It occurred to me I could also draw this out into a larger summer sermon series that included other fabrics in the worship space. Rather than focus solely on the issue of the flags – and possibly focus too much anxiety and heated feelings on the one topic – why not put that topic into the context of a larger series on what our fabrics say about whom we are as a church and what we are as God's people called to be servants of God's mission in the world?

This got me thinking about other fabrics in church:

Flags: Some churches include the U.S. flag as well as a Christian flag and perhaps a United Nations flag. Where these are located should reflect the U.S. Flag Code as set by Congress. If a choice is made to have them on display, then further choices include whether these are on the platform or the floor level. The U.S. flag is to be to the congregation's right side on the floor level. When the flag is on the platform or stage, however, the U.S. flag is to be to the speaker's right side and to the congregation's left.

Banners: What banners are hung, where, and when? Is there a liturgical focus to these banners, and what is the relationship of color, symbol, and words or phrases on these banners to our worship?

Drapes/draperies/curtains: Where might these be found in the sanctuary? Why are they there? Do the colors hold any significance? Is there any ceremonial or worship function to these fabrics? Is there a shading/lighting function they fulfill, or do they elicit a sense of mystery – what is behind the curtain, and should we know?

Altar cloths and paraments: Why do we place cloths on communion tables and altars? What is the significance of the colors and any symbols that are on these fabrics? What do these say about our worship and liturgy? What fabrics hang from the lectern and pulpit? These forms say something about our worship, and they have a context, or story, and then a number of meanings we can draw.

Vestments: What do the clergy wear while leading worship, and is this important? What are the variety of robes, albs, and chasubles that are worn? What is the history of these garments? What do the styles and colors have to say about the role and function of the clergy in our worship? What about stoles and their colors and symbols? What is their story, and what meanings do we draw from them? Are there trends away from wearing these formal vestments? If the clergy wear street clothes, is this appropriate, and what might this be telling us?

Dress codes: Does our church have a dress code for worship? Is the code clearly spelled out, or is it unspoken? What should we wear? Do we carry expectations? Standards? Anything goes? Does it matter, really? Is this essential or non-essential? Naming this subject is an important one and will elicit lots of response from church members, since it's a subject that's possibly addressed privately but seldom publicly. It would help to get these issues out on the table!

As part of my sermon series I developed a PowerPoint presentation that showed a number of people wearing different clothing, jewelry, and tattoos and invited the congregation's response as part of the sermon/message. I asked: Would this person be welcome in our church? Would this one? I also showed pictures of guide dogs and people in wheelchairs. The combined presentation on clothing and differing looks and abilities opened up the larger issue of who is really welcome to our worship.

In the next chapter I provide a model for preaching with your windows.

Chapter 12

Preaching Your Windows

The sanctuary windows are an important structural and decorative part of the church. Whether they are clear or stained glass, they deserve at least one if not a whole series of sermons.

There is a long and interesting history on the structural challenges involved in building massive church walls and including long sections of windows: One need look only to Notre Dame in Paris and the flying buttresses to see one solution to the problem of supporting heavy walls with holes in them: windows.

Other French churches such as St. Chapelle in Paris, St. Denis just outside of Paris, and the Cathedral at Chartres offer wonderful stories and experiences about the development of stained glass.

Sermons connecting this history with the history of one's own church building can be interesting, particularly as stories are told. Some of them are found in the windows themselves, with their stories, symbols, and colors, and how these are connected to church history and theology. Some stories are found in the names of persons or groups that gave windows as memorials and learning whom they were and what prompted the gifts.

For those churches without stained glass, there is also a history as to why they opted away from that, sometimes rooted in the iconoclastic movements of the Protestant Reformation and a commitment to simplicity, a suspicion of visual imagery, and a prejudice toward the preached/spoken word. Other times the presence of clear glass, particularly in more natural rural or suburban settings, serves to open the sanctuary to light and to the nature surrounding the church building.

Some windows are designed to be opened to bring in fresh air or to let out the sound of organ and singing. These functions also serve meanings that may be drawn out.

As you sit in the sanctuary you might notice how many windows there are on a wall: Does the number bear any significance? How many panes are in the windows? Does this number bear any significance? I have often noticed sets of six, seven, eight, and 12 in church windows, all non-verbally speaking to symbolisms we've already noticed in previous chapters.

As part of your studied observations with your sanctuary, you might notice how the light changes at different times of the day and how what the windows bring in changes with that light. How do different weather conditions affect the light – and mood – thrown by those windows? Is there a window with imagery in the front of the sanctuary, and what does that say to the gathered congregation? Is there a window with a story that all see as they enter or leave the narthex or the nave? What significance does that window and its story have for the worship congregation and worship itself?

For example, one church I served had a large stained glass picture of Jesus and a flock of sheep, and he was holding one sheep, presumably a reference to the parable of the lost sheep. This window was only seen as people left the worship sanctuary. When that church building was sold and a new one built, that window was moved to the new church. This time, the window was placed in the chancel area, where the window was seen throughout the entire worship service.

Rather than being a focus of leaving, representing how the congregation was sent out into the world to find comfort that Jesus seeks the lost, and that they too might follow him and do the same, the same imagery became a part of the worship experience. Did anything change? Was there a new emphasis that the worshipper might find comfort and belonging during worship but might not make a connection with how he or she is to also look for lost sheep in the world outside the sanctuary?

In other words, one placement of the window invited an outward response as the people left the church to return to the world, while the other placement of the window invited an inward response as people gathered for worship to be fed and found. Or not. The form is the stained glass window, the content is the story it tells, and the meaning, the "so what?" is in the eyes and heart of the interpreters, which means there is no one answer!

Preaching your windows involves spending time with each one. Many churches have individual stories represented in each window, shown as pictures, symbols, and colors. The pictures may depict references to biblical stories from the life of Jesus (lost sheep, knocking on a door, rescuing Peter from the water, at prayer), sacraments like baptism or communion, or even stories from the life of the congregation gathered in that church.

The pictures might be combined with symbols such as loaves of bread, sheaves of wheat, grapes, water, doves, flames, vines, light rays, or with historical Greek and Latin symbols such as the Alpha and Omega, the Chi Rho, or the INRI.

"If churches are made radiant and beautiful places of worship, we can have a spiritual regeneration without anyone knowing what is going on. Beauty can preach as very few men with bundles of words can preach. I want to make beautiful interiors for both churches and souls. I want men (*sic*) to hear my windows singing; to hear them singing of God. I want men (*sic*) to know that God is at the core of their own souls."

Charles J. Connick, whose studios made stained glass windows for churches for nearly 75 years

Stained glass windows contain a rich and colorful lexicon of biblical story and theological affirmation people have seen time and again and yet have never really noticed, much less understood the references. Preaching your windows invites the congregation into a new way of seeing their home sanctuary, and even visitors will get something out of this to take back to other sanctuaries they may visit.

Inviting people to look at the windows as you speak also offers an audiovisual presentation without needing any additional technology!

It has been my experience that people report they've worshipped in a sanctuary their entire lives and never noticed what was going on in those windows – and how people will always look at the windows differently from now on. That makes for a memorable sermon!

What follows is an example of how you could begin to reflect out loud in a sermon about the windows in your sanctuary. We can trust that my reference to south-side windows actually fits your sanctuary's orientation, but you can make your own adaptations.

What a wonderful testimony that you have put these memorial windows on the south side of the church. They face the side that receives the most morning light. How significant it is that they remind us of the light that God made in the beginning when all was dark and formless and the spirit of God hovered over it all and God said, "Let there be light!" There was light, that first light at the beginning, but more than that now, these windows bring in a light meant for us to enjoy on Sunday mornings. So, the windows also represent the first Easter morning and the light of resurrection faith.

The windows bring us the golden light of the fresh new morning, and they also bring us rich and vivid colors. You may not even be aware of this, because each contrasting color, as different and bold as it is, still serves the overall harmony and beauty that are ever gifts from God.

The great 20th century painter Wassily Kandinsky said that " . . . colour is a power which directly influences the soul. " He wrote a little book called *Concerning the Spiritual in Art.* (You might wish to buy his book and read directly from it as you preach your message.) He was fascinated with how colors reached into human souls with the complete and full power of the spiritual. He said colors produce a "spiritual vibration."

Look at the colors in the window, if you like, while I read to you how he describes some of the colors in the window and their spiritual flavor.

"Blue is the typical heavenly colour. The ultimate feeling it creates is one of rest." (Note, too, that we associate the color blue with water as well, and this color in your windows can refer to the waters of life, the waters of chaos out of which God brought life [see Gen. 1:1])

"Green is the most restful colour that exists . . . (I)t is the colour of summer, the period when nature is resting from the storms of winter and the productive energy of spring." (Green can refer to nature, life, growth, and even eternity.)

"Red . . . rings inwardly with a determined and powerful intensity." (Red, in the church, also refers to the Holy Spirit, or to renewal, in particular as found in the Pentecost story in Acts 2.)

The brown shows a "powerful inner harmony . . . an appeal of extraordinary, indescribable beauty." (It can also refer to the color of earth, of ground and soil, the foundation of life and agriculture.)

Yellow can refer to holiness, as God-light dispersed into human hearts or shaped as a halo over heads of Jesus, the disciples, angels, and other holy figures.

"Shades of colour," he wrote, "awake in the soul emotions too fine to be expressed in words."

Today we celebrate how we are surrounded by so great a cloud of witnesses who bring us testimony to the power of faith, hope, and love. These windows, indeed all the windows in this sanctuary, touch our souls with delicacy, beauty, and light. They speak of the relationship of a solid structure (the wall itself) and of a necessary openness (the windows). Together wall and window tell us something of the nature of the church: a vessel by which we bear witness to fragile grace, and to enlightening beauty.

Today we give thanks for those who have given us this beauty and who invite us, each time we gather in this sanctuary, to testify to God's abundant love for us.

These windows (in the case of memorial windows) could represent the losses of the past. The windows could remind us of the passing of a beloved minister; they could remind us of the loss of a beloved old church building; and of the losses of church members and servants of this church. But the beauty of these windows, the Easter faith to which they testify, best serves as a moving forward in confidence, a building of a healthy new congregation strong and vital, ready to greet the future with beauty, grace, and color.

If this is so, then it is time to turn the shadows of loss into the brightness of God's colorful and golden future. Let these windows stand for the faith, hope, and love of those who have gone on before us and who have dedicated their lives to this church. Let these windows be our comfort, our guide, and our inspiration as we rededicate ourselves to the uniqueness that is the (your church name here).

For as the writer of the Book of Hebrews (11:40-12:1) declares: " . . . since God had provided something better so that they (the dear ones we remember) would not, apart from us, be made perfect. Therefore, since we are surrounded by so great a cloud of witnesses, let us also lay aside every weight, and the sin which clings so closely, and let us run with perseverance the race that is set before us."

As we close these three chapters on the nave and chancel, we move into some additional preaching opportunities with architecture in the following chapter.

Note

The quotations in the discussion on colors are from Wassily Kandinsky, *Concerning the Spiritual in Art* (New York: Dover Publications, 1977), pp. 25 and 38.

Chapter 13

Additional Preaching Suggestions

In this chapter we'll look at a few other possibilities for preaching with illustrations from the rich visual library that is your church. The first example is based on the outside of your church building. The second example features a look at some of the artwork on display inside the church. We close with some thoughts on how all this can be pulled together into a series of sermons.

Preaching the Exterior

As part of your walk around the church, you might spend some time outside. Is the church site on a hill or a level surface? Is it a high point (elevation) in the community? Is the building raised up on steps? The theological significance of this, "going up" as the Psalmist writes in Psalms of Ascent as pilgrims walked up to Jerusalem, and symbolically going upward and heavenward as a spiritual pilgrimage, is often a nonverbal message found in a staircase or set of steps leading up to a church building. Is this theological commitment borne out in older architecture now an impediment to those who in today's world need more accessible facilities? Has the church accommodated these needs with ramps and/or elevators or an alternative entrance to help people access your space?

What is the shape of the exterior? How many exterior walls are there? How would you describe the general shape of the building? Does it follow a cross pattern? Does it look more like a ship or a boat?

How do the windows look from the outside? If they are stained glass, you don't see as much as you see from the inside. This in itself invites a meditation on stained glass windows and the "window" of our soul. People might look at those exterior windows and know they are windows but little else about them until people get inside and see the colors and stories. So, too, this might be how we appear: We see people on the outside but have little access to the full color and story of their lives, their inner worlds. These observations might round out the messages shaped around your windows as well (see previous chapter).

What else is there outside the building in the way of statues, installations, bell towers, crosses, artworks, plants, shrubs, and trees? How effective is your signage? Can people driving by see your sign and get information about your church they might be seeking? Do these connect with your church's message and mission? Are they inviting? What do they say about whom this congregation is and who is welcome?

What are the forms of your exterior, what is their story, how does that story connect to what goes on in your place, and what are some of the meanings of what is seen? Bringing this to the attention of the whole congregation in the context of the larger Christian story could make an interesting sermon series. Since your congregation is inside when the message is offered (unless you do offer this sermon during an outdoor worship where you are seated outside the building), you might wish to take pictures for the main preaching points and put them into a presentation program such as PowerPoint and project the photographs on a screen in the sanctuary so all can see what you've noticed from the church exterior.

Preaching With Your Church's Art

Most churches have popular religious paintings hanging on the walls. Some of these are stories from the Gospel accounts of the life of Jesus. Popular images include Sallman's "Head of Christ" (1941) and various versions of Leonardo da Vinci's "The Last Supper" (1495-98).

As with everything else in the church, many people walk by these pictures and give them little notice, while some people find comfort and meaning knowing the pictures are there.

One way to preach these pictures is to find an easel, and if they are large enough, bring them into the sanctuary so people can see them as you talk about their content and meaning. Alternatively you could find digital images of these same pictures on the Internet and project them on a projection screen so people can see a large image and even close-ups of details of the picture.

A sermon could include looking at the picture, identifying the particular biblical passage the picture illustrates, speak to what's known about the artist and the artist's commitment to the subject, and what meanings may be drawn for the life of the gathered congregation that day.

Since there are so many "Last Supper" pictures that refer to Leonardo's version, I have found that people enjoy seeing what other artists have done side by side and then comparing them with the original. This technique opens a lesson in interpretation and how artists preach their own vision of messages of the story in subtle and sometimes not so obvious ways. If nothing else, this teaches people how to look more closely at art and receive permission to study art and draw their own conclusions.

"A church living aesthetically will find itself breaking new ground. There will be a cessation of old, desiccated ways and, perchance, an ecstatic movement toward freedom."

John Westerhoff

I have preached this way and found people appreciate learning more about the art and artists that have been a part of people's churches for many years.

Another growing point for a congregation is to introduce it to other artists' depictions of these common biblical stories such as the Last Supper, the parable of the lost sheep, Christ knocking at the door, or the head of Christ. When done from a cross-cultural experience, and by bringing in the work of artists from Asia, Latin America, and Africa, a congregation's sense of global Christianity will grow. This will also show how a congregation's bias toward northern European art can limit the congregation's visual horizon.

The church I mentioned in the previous chapter that had the large stained glass image of Jesus with the lost sheep front and center for the congregation hadn't paid much attention to the fact that Jesus appears as a very white male. The other stained glass windows also showed Jesus as white European, as did the rest of the art around the church. Since the Sunday school was multiracial, it was essential to me to start bringing to the congregation the rich multicultural imagery of Jesus from around the world.

Using the blank white wall to the right of the pulpit as a projection screen, I developed regular PowerPoint presentations to show such things as the nativity of Christ from multicultural perspectives or some of the parables of Jesus. For our monthly communion Sundays we would show a variety of "last supper" images or of people eating together in different global settings. In a way this "wall" with the projected imagery became another stained glass window showing other stories from the life of Christ from varied sources.

On Doing A Sermon Series

Much of what we've discussed here can't be preached in just one sermon. The subject matter would take a series of sermons to effectively handle the vast amount of material touched upon in these pages. Besides being able to cover more about the architecture, furnishings, art, and fabrics of your church, a sermon series also seems to keep people attending.

Having a series is easy to promote, and a series such as this is unique and might well draw visitors. The preacher enjoys benefits from a series as well, as it's easy to structure and sets the way ahead for some weeks and even months. The series helps structure material and focus messages around a theme, heightening a sense of organization that sometimes is lost with the seeming randomness of lectionary preaching alone.

Summers are a good time for a sermon series of maybe 3-4 weeks. Sermon series are good in the summer because, if they are interesting enough to people, they will boost attendance. People might make an effort to attend all three or four weeks of your series.

That said, the fall is another good time for a sermon series when people are returning to church and when, if they are returning after a busy summer, they might respond well to fresh reminders of their identity as members of a particular congregation of the Body of Christ in their part of the world.

Certainly the topics I've discussed may be preached as stand-alone topics, but these topics could also be folded into other sermons, particularly when references may be made to something in the sanctuary that relates to your topic. For example, if you're preaching a Communion sermon, you could refer to the table itself and what it signifies, or you could refer to the words on the table, "This Do In Remembrance of Me," or to the Alpha and Omega that might also be carved into that table.

A sermon on a parable might also point to that same parable that happens to be illustrated in one of your windows. A Good Friday sermon could refer to the INRI found on a banner in the sanctuary or to the cross pattern (Greek or Latin) upon which the congregation gathers as it is seated in that nave.

Long ago theologian Paul Tillich reflected on how, over time, religious symbols can lose their meaning and become simply an unseen background.

As many churches struggle with their own relevance in a quickly changing world, describing the symbols, art, and architecture of one's worship space may just be enough to inspire congregations to move forward with new energy.

Chapter 14

Exercises

In this chapter we will look at pictures from several worship sanctuaries and put to practice some of the principles described in the chapters in Part III. Using a form-content-meaning method, I will draw out information from each photograph, and invite you to do the same. We'll end each example with some possible preaching directions for each picture.

A Method for Interacting With A Worship Space

Form: what we see: structural features, furnishings, fabrics, symbols.

Content: The stories told and references made: to biblical stories, to theological symbols, to a congregation's history, to a particular worship event.

Meaning: Possible interpretations and lessons drawn from the forms and content.

Preaching possibilities: Suggestions for drawing a congregation's attention to the forms, stories, and meanings of a worship space.

Form: The photograph is of stained glass windows on the west wall of St. John's United Church of Christ in Hartford, Wisconsin. There are eight sections of windows framed with wood. There are four long windows, two four-sided kite-like framed shapes at the top of each pair of long windows, and topping it all off is a circle with twelve panes, beneath which is a small triangle. Each of the four long windows has eight sections starting with a bottom rectangular section, above which are six smaller rectangles. The other panels include oval shapes, with a symbol in the fourth section of panels in each of the four long vertical sets of windows.

Content: Historical records show this sanctuary was built in 1907, and we might assume the windows were installed at that time. Colors include pale white, brown, green, and pink. In the fourth panel counting from the bottom is a representation of a *fleur-de-lis*, a lily. There is a bright light on the upper left side of these windows coming from sunshine through similar windows directly across on the east wall.

Meaning: Numbers: The round circle at the top could represent the world or a circle of eternity or a clock face with twelve panels – twelve hours? twelve months in a year? twelve tribes? twelve disciples? The four-sided diamond shapes could be three triangles divided in half (Trinitarian symbolism), and each includes eight panels (six represents creation of heaven and earth, seven represents rest and worship, eight might represent a transcending beyond the physical and spiritual and a fresh beginning). Overall, there are eight large sections framed by the wood, and each pair of windows has eight ovals in each horizontal section. The rectangular base could be the foundation, and the six smaller rectangular brick-like panels could be the structure of earth/life resting upon the seventh day as the day of rest and worship.

The three petals of the lily may represent the Holy Trinity, with the horizontal band at the middle symbolizing Mary, the mother of Jesus. The flower might also symbolize both the Annunciation to Mary and the Resurrection of Christ, and the ovals tucked beneath the drooping flower stalks might represent joy, renewal, or unconditional love. The earthy browns and vivid greens might recall soil and plants, the natural world of a religious community gathered to worship.

St. John's United Church of Christ, Hartford, Wisconsin

Preaching possibilities: Draw out meanings of light coming into the sanctuary; the number symbolisms of six, seven, and eight and what that means for a worship community; the passage of time with the clock-like window at the top, the fresh beginning of a week opening with worship; the meanings of the lily with its many references to biblical stories. These windows do more than bring a glorious light into the worship sanctuary: they feed imaginations and open hearts and minds to spiritual beauty.

Form: The photograph is of the chancel area of Union-Congregational UCC in Waupun, Wisconsin. Three steps lead to the chancel with a choir at the upper left, with a window with six panels, a draped cross on the back wall, a communion table or altar with a book on the table and a candle in front of the table, resting upon fabric reading "Holy Holy Holy." Front left is a smaller lectern, and to the right is a larger pulpit, each with fabric and a symbol of a crown. Behind the pulpit is a projection screen with fabric extending below. An unseen projector is behind the screen. On the floor level (the nave) are three rows of armed chairs in the front and behind are pews.

Content: A color image would show purple cloths, so the church is presumably decorated for Advent or Lent. The simplicity of the space without further decoration might suggest Lent, as would the draping of the cross. Three steps up might recall the Holy Trinity; the "Holy Holy Holy" reference is to Isaiah 6:3 or to the hymn of the same name. The crown could be a reference to the crown of thorns put on Jesus' head by Roman soldiers in John 19:2 or to the hymn "Crown Him With Many Crowns." The screen is anchored to the right wall and is behind the pulpit. The screen is ready to show words, pictures, or both and bring more content into the room. There is an invisible line from cross to the pulpit and from cross to the lectern, showing a relationship between the three.

Meaning: What do the colors, symbols, and words of Lent mean for a gathered congregation? What does it mean that the screen is located where it is? How do the furnishings covered with fabric connect with one another, and what might that mean for the worship life of a congregation gathered in that space?

Preaching possibilities: One sermon could focus on the fabrics and their connection (color, words, symbols, biblical references) to the Lenten story. Another sermon could focus on the chancel area as a place set apart from the nave and what function that serves in worship and the life of the church. For example, one could reflect on the function of the people who lead worship from the chancel (pastor, choir, liturgists, etc.). Yet another could focus on the purpose and function of the relationship of cross, altar/table, lectern, and pulpit. One could talk about the location of the screen and why it is

Union-Congregational UCC, Waupun, Wisconsin

located behind the pulpit. While a wall serves as a convenient anchor, the screen is also positioned behind the place in the chancel where the word is preached: the pulpit. Since we "read" from left to right, the preacher might best stand at the lectern or in the middle of the chancel when imagery is posted on the screen. Noting the distance between lectern, where scripture is read, and the pulpit, where the word is preached, one can talk about the interpretive tension between the two.

Form: The photograph is of the sanctuary of Ozark Prairie Presbyterian Church in Mt. Vernon, Missouri. Looking at the front from left to right we see:

An organ, a woman behind a small lectern with a white cloth on it, a U.S. flag, a man standing at a pulpit with white cloth on it, three chairs with cross-like structures, a large picture of a bearded man in a white robe draped with a colored cloth and arms outstretched, a slightly crooked projected image, below the image is a candleabra with seven candles, in front of this is a table or altar with a floral arrangement; visible beneath are three steps below, and off to the right is a cross topped with a thorny crown and draped with a white cloth, with a church flag to the right. Below the cross is a small projector on a stack of books, and just visible to the right of that is a computer screen. Looking up we see three light fixtures, and a ceiling line evoking a triangle.

Content: This is during a denominational business meeting. The central picture is Jesus in a welcoming pose. The projected image uses a light-colored wall for adequate contrast and visibility. The U.S. flag is properly situated at the speaker's right on the platform. The white cloths are seasonal in a church, usually used at Christmas and Easter, and with the draped cross and crown of thorns we could presume the Easter season.

Meaning: The chancel is three steps up, showing a Trinitarian symbolism, and there are three chairs as well. The tallest chair might be for the honored speaker or preacher. The cross represents the crucifixion of Jesus, and the thorny crown shows suffering even as the white cloth (the burial shroud?) evokes resurrection and John 20:6. What does it mean that of the two flags present, the U.S. flag and the Christian flag, that the U.S. flag is very close to the pulpit? Is this for a reason?

Preaching possibilities: One sermon could focus on the Easter themes with the white cloths. Another could focus on questions raised and statements made by the positioning of the flags. Another message could focus on the nature of Calvinistic Presbyterianism and its iconoclastic history and suspicion of imagery in worship in relation to all the imagery in the chancel. What is the history of the Jesus picture in that sanctuary and how does it match up with or conflict with an anti-imagery tradition? Does this invite using projected images in worship where once this would never be considered?

Ozark Prairie Presbyterian Church, Mt. Vernon, Missouri

Form: This photograph shows the nave and chancel area of the former University Christian Church in Berkeley, California. The church provides a very unique example of the imaginative use of fabric in a sanctuary. What is shown is a temporary fabric installation that implies lightness and delicacy while capturing visual attention.

Content: The church is hosting a special lecture event sponsored by the Pacific School of Religion in 2010, "Spiritual But Not Religious: Chasing the Divine." We see a ceiling with wooden rafters, below which hangs a translucent fabric as a kind of canopy inside the worship space. The separate arms of the fabric soar upward. At the left edge the fabric is shaped as a screen for a projected image. There is a single stained glass window at the front wall with what looks to be a cross shape. In front of that is another temporary installation that involves colored panels.

Meaning: Do the wooden rafters seem like an ark-like structure or more barn-like? What feelings does the natural wood evoke? Do the rafters provide structural stability and more of a visual backdrop to the space than anything symbolic? There is a thematic connection with the hand imagery on the central colored panels repeated with the image on the fabric screen. The hand has a hole in it, like a sun with beams of light. There is a sense of both immanence and transcendence in the hand: practical "hands-on" work in the world, joined with meditation and prayer. Note how the white fabric serves as a kind of canopy that captures attention while also functioning as a screen inviting an imaginative aesthetic.

Preaching possibilities: What do the fabrics evoke as they soar upward? How does the interaction of permanent structure, temporary installation, color, fabric, symbol, and light draw out the theme of "Chasing the Divine"? With so much going on visually, how does the spoken word interact with the visual "word" as presented? A preacher could engage the congregation by asking questions about its experience with the space. What does it look like from the congregation's perspective? What does the space feel like? How does it fit the theme?

University Christian Church, Berkeley, California

Form: This photograph shows the Cocoa Beach Community Church in Cocoa Beach, Florida, immediately following a Sunday worship service.

Content: There is a ceiling with wooden rafters. At the front are three stained glass windows and two flat screen televisions to either side. Light fixtures seem to be in groups of three. There is a U.S. flag off to the left on the platform, but the flag is not prominently displayed. To the right of it is an organ, with a baptismal font on the floor in front of the organ, and in the center is an altar or communion table. The pulpit is on the far right side behind a wooden box.

Meaning: Do the wooden rafters seem more barn-like, remembering the birth of Jesus in a stable? What feelings does the natural wood evoke? There are triangular shapes in the ceiling, as well as three steps leading up to the chancel, invoking Trinitarian symbolism. There is a visual balance in the chancel with the three stained glass windows flanked by the two flat screens. The angle of the walls seems to indicate an octagonal shape, with three walls in the front, long side walls to the back, and three angled walls at the rear. The eight sides, as noted before, might remind the community of how worship in this space calls forth fresh beginnings in a new day.

Preaching possibilities: The three windows and two screens provide a group of five referring to the first five books of the Pentateuch or Torah, used for instruction. The three-panel central stained glass window and the two on the side, while not clear to us in this picture, yet have stories to tell that could be preached. During worship the flat screens were used to show hymns to songs and prayers and to post announcements after the service was finished. The screens were turned off during the sermon. They could be used to add additional imagery to a sermon. For example, during a sermon on the stained glass windows, close-up images of certain features of the windows could be displayed on the screens for people to get a better view. Given the three windows, the lights in groups of three, and the rafters in triangles, a sermon on the meaning of such Trinitarian symbolism for that particular congregation might prove interesting to them.

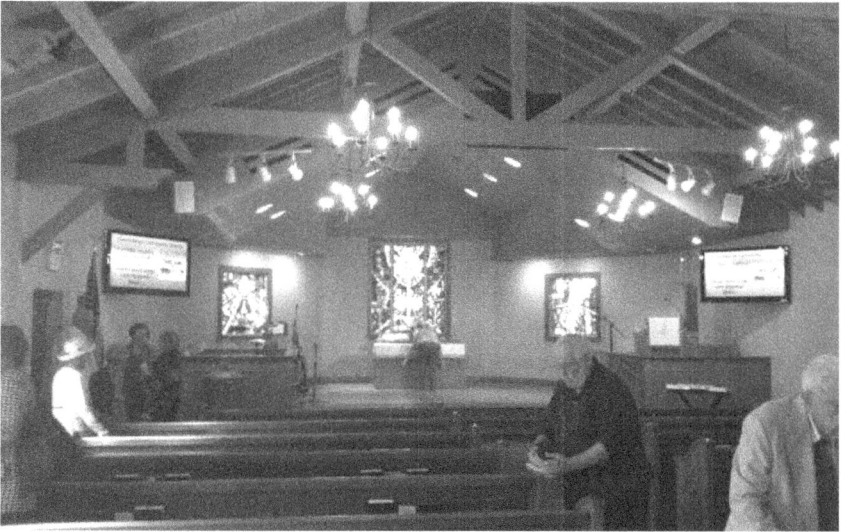

Cocoa Beach Community Church, Cocoa Beach, Florida

Form: This photograph is taken of St. John's United Church of Christ in Hartford, Wisconsin. At the front on the floor level is a table with a white cloth. Two steps up is the platform, where on the right is a lectern with an open book and a symbol or pattern of three overlapping ovals and on the left is a partial view of a pulpit with a similar cloth. A baptismal font is directly in front of the pulpit. An organ is behind the pulpit and a bench behind the lectern. Up three more steps is the chancel, marked by the wooden railings on either side, and at the top is an altar with a cloth, a Bible, a floral arrangement, and two candles. Centered behind is a Latin cross. The wooden screen behind the altar is, like the windows from the same church, using groups of four panels. To the top above the cross there are five niches and a small carved cross on either side. Below each of the small crosses is a line of seven small squares. Two banners flank this screen, each with a picture of a candle, and one with the word "Jesus" and the other with the word "joy."

Content: The table on the floor level is a communion table, signaled by the words, partially obscured by bright sunshine flooding the table, "This Do In Remembrance of Me" from I Corinthians 11:24. This would make the furnishing directly beneath the central cross an altar. The communion table would hold the communion elements, while the altar would be the place to present offerings.

Meaning: There are a cross below the communion table and two shrine-like openings on either side. The cross has the symbol IHS on it, Greek letters for the words Jesus, Son, Savior. The symbol on the visible altar cloths looks like a summary of the *IXTHUS*, a Greek anagram for "Jesus Christ, God's Son, Savior." When the letters are drawn together in a circle, two patterns may be seen: a Greek cross and what looks to the modern eye to be an airplane propeller. The "propeller" image on the altar cloth may represent this shape. As it is now it may evoke the Trinity. The flames of the altar candles might represent the light of God's presence during worship, and the flames on the banners might represent Jesus as light of the world (John 8:12). On the large wooden piece behind the altar the sets of four panels add to eight across and follow the "eighth day" motif as a fresh beginning. The seven small squares might represent the six days of creation and the seventh day of rest and worship, and the five niches at the top might represent the first five books of Moses in the Old Testament. Three niches on the front of the altar speak of

St. John's United Church of Christ, Hartford, Wisconsin

Trinitarian symbolism. Pews visible in the chancel are to be used by the choir.

Preaching possibilities: A sermon could discuss the difference between a communion table and an altar. The separate chancel invites discussion of what is held most sacred at the altar area: cross, Bible, Trinity, an altar for offerings, and a place for the choir. Why are these given a favored position, and what does that mean biblically and theologically?

Form: This is a photograph of one stained glass window in a set of eight at St. John's United Church of Christ in Slinger, Wisconsin. The windows were dedicated in 1950 in the original church building and later moved to the new facility.

Content: A picture is set in an oval shape. Along the sides of the whole window are what seem to be curling branches with leaves, and above and below the oval are flowers and leaves. Two curling arrows may be seen on either side of the oval, pointing towards the oval. The picture in the oval is of a star with three different kinds of light beams, one set circling the star itself, another pointing outward, and the third set flowing downward to a small covered area. The star seems to be beneath a dome with curling smoke or clouds just above the roof. Light rays seem to point to a flower on a bed of hay resting in a feeding trough or manger, below which is the stem of the plant with hay or ground on either side.

Meaning: The picture speaks to the birth of Jesus at Bethlehem (Luke 2:7) where the child is laid in a manger. A star shines overhead (Matthew 2:2). The rose may be from Song of Solomon 2:1, "I am the rose of Sharon . . . " sometimes applied to Jesus as perfect and beautiful and as one who loves. The leaves below the manger and below the oval seem more like those of the acanthus plant, a symbol of immortality.

Preaching possibilities: Using this window in a Christmas sermon could draw out the symbols associated with the birth of Jesus: star, manger, rose, acanthus leaves. There is also a cross pattern on the vertical from star to plant and the horizontal of the roof. Birth, death, and resurrection are found in this window, recalling lines from the third verse of Charles Wesley's hymn "Hark! the Herald Angels Sing:" "Hail the heaven born Prince of Peace! Hail the Son of righteousness, Light and life to all he brings, risen with healing in his wings."

St. John's United Church of Christ, Slinger, Wisconsin

I extend my thanks to the following churches for inviting me into their spaces, for what I have learned from them, and for the photographed images included in this chapter:

Cocoa Beach Community Church UCC, Cocoa Beach, Florida
Epworth United Methodist Church, Berkeley, California
Ozark Prairie Presbyterian "Brick" Church, Mt. Vernon, Missouri
St. John's United Church of Christ, Hartford, Wisconsin
St. John's United Church of Christ, Slinger, Wisconsin
St. Luke's United Methodist Church, Dubuque, Iowa
Union-Congregational Church UCC, Waupun, Wisconsin
University Christian Church, Berkeley, California.

About the Highlighted Quotations

Opening – Albert Einstein, *Cosmic Religion: With Other Opinions and Aphorisms*, 1931.

Introduction – Pope John Paul II, *Letter to Artists,* Easter Sunday, April 4, 1999.

Chapter 1 – Pope John Paul II, *Cinema: Communicator of Culture and Values,* Message for the 1995 World Day of Communications, May 28.

Chapter 2 – Walter Ong, *Orality and Literacy: The Technologizing of the Word (*Methuen & Co. Ltd. 1982).

Chapter 3 – Camille Paglia, *The Magic of Images: Word and Picture in a Media Age* in *Arion* 11.3 Winter 2004.

Chapter 4 – John H. Westerhoff III , *Will Our Children Have Faith?* (Harrisburg, Pennsylvania: Morehouse Publishing, 2000).

Chapter 5 – Kent C. Bloomer and Charles W. Moore, *Body, Memory, and Architecture* (New Haven: Yale University Press, 1977).

Chapter 6 – Marshall McLuhan, *The Medium and the Light: Reflections on Religion and Media* (Toronto: Stoddard, 1999).

Chapter 7 – Marshall McLuhan, *The Medium and the Light: Reflections on Religion and Media* (Toronto: Stoddard, 1999).

Chapter 8 – Kim Veltman*, Understanding New Media: Augmented Knowledge and Culture* (University of Calgary Press, 2006).

Chapter 9 – Michael Bausch

Chapter 10 – Kent C. Bloomer and Charles W. Moore, *Body, Memory, and Architecture* (New Haven: Yale University Press, 1977).

Chapter 11 – Charles J. Connick, whose studios made stained glass windows for churches for nearly 75 years in the 20th century.

Chapter 12 – John H. Westerhoff III, *Will Our Children Have Faith?* (Harrisburg, Pennsylvania: Morehouse Publishing, 2000).

Chapter 13 – Michael Bausch

About the Author

The Rev. Michael Bausch, D. Min., is a parish minister, author, and educator who teaches the use of the arts and multimedia in worship based on many years of parish ministry experience in the United Church of Christ. Over the last fifteen years he has been a keynote speaker at a large number of UCC and Presbyterian denominational events.

His teaching experience includes offering summer courses in the subject for several years at the Graduate Theological Union in Berkeley and in a number of other mainline seminaries, including the University of Dubuque Theological Seminary, the Vancouver School of Theology, United Theological Seminary of the Twin Cities, and Luther Seminary. Besides teaching biblical studies courses at the University of Wisconsin-Platteville, he has extensive online teaching experience that includes developing and teaching courses at undergraduate and graduate levels, including summer online courses for the Pacific School of Religion.

His books include *A Media Sourcebook* (Pacific School of Religion, 1973), *Everflowing Streams: Songs for Worship* (with Ruth Duck, The Pilgrim Press, 1981), and *Silver Screen, Sacred Story: Using Multimedia in Worship* (The Alban Institute, 2002). He has written numerous articles for publications such as *The Clergy Journal, Liturgy: Journal of the Liturgical Conference,* and *Church Educator.*

Presently he is serving as an interim minister in the Wisconsin Conference of the United Church of Christ. Other interests include playing guitar and harmonica with several musical groups, playing tennis with friends, enjoying family life with his wife Cathy and their two grown daughters and son-in-law, and horsing around and doing art projects with his grandchildren.

About the Publisher

Image Source, which is owned and operated by Fred Noer, provides writing and editing services and black-and-white photographic images of landscapes. For more information, go to the Web site www.frednoer.com.

www.ingramcontent.com/pod-product-compliance
Lightning Source LLC
Chambersburg PA
CBHW061728020426
42331CB00006B/1148